T0371601

Elements in the Philosophy of Ludwig Wittgenstein
edited by
David G. Stern
University of Iowa

WITTGENSTEIN AND RUSSELL: THE VICISSITUDES OF JUDGMENT AND PROPOSITION

Sanford Shieh
Wesleyan University

Shaftesbury Road, Cambridge CB2 8EA, United Kingdom

One Liberty Plaza, 20th Floor, New York, NY 10006, USA

477 Williamstown Road, Port Melbourne, VIC 3207, Australia

314–321, 3rd Floor, Plot 3, Splendor Forum, Jasola District Centre, New Delhi – 110025, India

103 Penang Road, #05–06/07, Visioncrest Commercial, Singapore 238467

Cambridge University Press is part of Cambridge University Press & Assessment, a department of the University of Cambridge.

We share the University's mission to contribute to society through the pursuit of education, learning and research at the highest international levels of excellence.

www.cambridge.org
Information on this title: www.cambridge.org/9781009468145

DOI: 10.1017/9781108946858

When citing this work, please include a reference to the DOI 10.1017/9781108946858

First published 2024

A catalogue record for this publication is available from the British Library.

ISBN 978-1-009-46814-5 Hardback
ISBN 978-1-108-92509-9 Paperback
ISSN 2632-7112 (online)
ISSN 2632-7104 (print)

Wittgenstein and Russell: The Vicissitudes of Judgment and Proposition

Elements in the Philosophy of Ludwig Wittgenstein

DOI: 10.1017/9781108946858
First published online: February 2024

Sanford Shieh
Wesleyan University

Author for correspondence: Sanford Shieh, sshieh@wesleyan.edu

Abstract: Responding to Russell is a constant throughout Wittgenstein's philosophizing. This Element focuses on Wittgenstein's criticisms of Russell's theories of judgment in the summer of 1913. Wittgenstein's response to these criticisms is of first-rate importance for his early philosophical development, setting the path to the conceptions of proposition and of logic in *Tractatus Logico-Philosophicus*. This Element also touches on further aspects of Wittgenstein's responses to Russell: the rejection of Russell's and Frege's logicisms in the *Tractatus*, the critique of Russell's causal-behavioristic philosophy of mind in Wittgenstein's "middle" period, the Russellian origins of notions of privacy dialectically treated in *Philosophical Investigations*, and the discussion of "surveyability" of mathematical proof in *Remarks on the Foundations of Mathematics*, which is, again, a response to Russellian logicism.

Keywords: judgment, truth, falsity, representation, proposition, belief

ISBNs: 9781009468145 (HB), 9781108925099 (PB), 9781108946858 (OC)
ISSNs: 2632-7112 (online), 2632-7104 (print)

Contents

Introduction

In 1911, Wittgenstein, perhaps on the advice of Frege, went to Cambridge to study with Russell.[1] By the following academic year, Russell regarded Wittgenstein as a collaborator in mathematical and philosophical logic (see McGuinness 1988 chapter 4; Monk 1990, chapter 3). What exactly they were working on together is uncertain. At least one project might have been a philosophical account of the nature of logic, to provide a foundation for the logic of *Principia Mathematica* (Whitehead and Russell 1910, hereafter *PM*; see McGuinness 1988, 104).

A crucial moment in their philosophical interaction was the summer of 1913, when Russell worked on a theory of judgment and understanding that Wittgenstein criticized at least twice. These criticisms marked the beginning of the end of their collaboration. They pursued diverging philosophical paths from then on. In the case of Wittgenstein, the summer 1913 criticisms of Russell played a, if not the most, significant role in Wittgenstein's philosophical development to the *Tractatus*.[2] In particular, they led to the conception of proposition (*Satz*) in that work. They are thus the principal foci of the present Element.

Section 1 presents the philosophical background and the basics of Russell's theory of judgment. Section 2 explores Wittgenstein's criticisms of it. Section 3 discusses Wittgenstein's initial theory of propositions intended to sidestep his criticisms of Russell. Within a year of formulating this theory, Wittgenstein came to doubt it because of difficulties recorded in a wartime notebook and elaborated in Section 4. Section 5 shows how these difficulties are overcome in the *Tractatus*. Ultimately, the Tractarian conception of proposition is a key basis of a very un-Russellian view of the nature of logic. This view is discussed in Section 6 along with Wittgenstein's differences with Russell's logicism.

The conception of logic in the *Tractatus* points to the broader philosophical-historical significance of the early interactions between Russell and Wittgenstein. Russell, along with Frege, rejected modal notions as philosophically unimportant. Frege asserted that "calling a statement necessary *has no meaning*" (1879, 5). Russell is more expansive: "the subject of modality ought to be banished from logic, since propositions are simply true or false, and there

[1] For uncertainty over whether it was Frege, see Monk (1990, 590).

[2] Citations of the *Tractatus* will be by remark number only; I largely follow the Ramsey–Ogden translation and a draft translation by Michael Beaney to whom I'm grateful for sharing with me. Occasionally I differ; the reader is advised to consult the original. That Ramsey played a major role in making the 1922 translation is established in Misak (2020).

is no such comparative and superlative of truth as is implied by the notions of contingency and necessity" (Russell 1905a, 520).[3] This banishment goes with a rejection of a long-standing Aristotelian modal characterization of the nature of logic: valid reasoning consists of conclusions following "out of necessity" from premises (*Analytica Priora*, 24b18–20). In the *Tractatus*, Wittgenstein brings modality back into logic: a primitive notion of possibility underlies the Tractarian conception of proposition, and the essence of logic lies in non arbitrary patterns of signification of all propositions.[4]

Sections 1–6 show the central importance, for Wittgenstein's early philosophy, of his grappling with Russell's ideas. It is no less important than Wittgenstein's early engagement with Frege's thought. Wittgenstein's responsiveness to Russell did not stop with the *Tractatus*. It is beyond the scope of this Element to chart in full the presences of Russell's philosophy throughout all periods of Wittgenstein's philosophizing. But, in Section 7, I touch very briefly on one later encounter with Russell's causal-behavioristic views of the mind that's continuous with Tractarian concerns, and indicate a couple of other later Wittgensteinian philosophical preoccupations that bear Russell's imprint.

1 Russell's Multiple-Relation Theories

1.1 The Rejection of Idealism and Moore–Russell Propositions

One source of the analytic tradition of philosophy is Moore and Russell's rejection of Bradley's absolute idealism. Moore and Russell had some common ground with Bradley. They all accepted that *judgment* is the same as *belief*, and that each judgment has an "*object*" (Moore 1901, 717). The basic intuition is this. If Socrates and Zeno both judge that Parmenides smiles, then they judge the same thing. It follows that there is an entity that they both judge. That entity is the object of their beliefs. The object of a judgment is a proposition. In the rest of this Element, I will also treat belief as interchangeable with judgment.

The disagreement is over the nature of propositions. Bradley conceives of propositions as formed from "ideal meanings" abstracted by cognitive activity from originally unified experience. Moore (1899, hereafter *NJ*) propounds arguments against the adequacy of this Bradleyan conception of judgment that

[3] In Shieh (2019), I elaborate the grounds and consequences of Frege's and Russell's anti-modal stance.

[4] The *Tractatus* is one of the two major returns of modality to logic in early analytic philosophy. The other is C. I. Lewis's critique of Russell's non-modal view of implication. I discuss Lewis in Shieh (2012, 2017, 2021, forthcoming).

motivate a radically anti-idealist theory of judgment.[5] Judgment and belief consist of *unmediated cognitive contact* with the world. Propositions are *not* thoughts or sentences, *not* mental or linguistic entities that *represent* purported facts in the world. Rather, propositions are complex entities whose constituents are parts of the world. If

Socrates believes that Parmenides smiles,

then the object of Socrates' belief is *a complex entity* composed of

- the person Parmenides, a part of the world, instantiating.
- the property or action or state of smiling, also a part of the world.

This complex entity is "about" Parmenides and "asserts" smiling of him (see Russell 1903 hereafter *PoM*, Section 43, chapter vii), so a Moore–Russell proposition *contains what it's about.* If representations are distinct from what's represented, then Moore–Russell propositions *don't represent* what they're about.

Two features of the Moore–Russell theory of propositions are particularly important for our purposes. First, Moore and Russell combine this theory with a rejection of a correspondence conception of truth. Truth is not correspondence to reality, but rather an indefinable property of propositions. The same holds for falsehood; it is not the absence of a corresponding aspect of reality, but a primitive property of propositions. Second, truth is fundamental; other notions are explained in terms of truth. In particular the notion of fact, or of the obtaining of a state of affairs in reality, is explained in terms of truth: "a fact appears to be merely a true proposition" (Russell 1904, 523). This is an anti-idealist conception of fact, for having a *true belief* is being in *direct contact with a fact* in the world.

A crucial consequence of these views of truth and of fact, drawn in *NJ*, is that there is no relativization of truth to space, time, or circumstance. In Russell's hands, this leads to the thoroughgoing anti-modal stance noted in the Introduction.

1.2 The Problem of False Propositions

From 1906 onward, Russell became increasingly suspicious of the Moore–Russell conception of propositions, especially its account of false propositions as unified complex entities with a primitive property of falsity. There are a

[5] Moore's arguments against Bradley have not, in general, been found very persuasive; however, in Shieh (2019, chapter 6), I argue that they present a cogent challenge to Bradley.

number of reasons for Russell's disquiet, but I focus on one explicit in "On the Nature of Truth and Falsehood" (Russell 1910, hereafter *NTF*).[6] Consider a purported false proposition such as the one expressed by

The Sun revolves around the Earth.

This consists of the relation *revolve around* uniting the Sun to the Earth in a single entity that has the unanalyzable property of falsehood. But if the *revolve around* relation unites the Sun to the Earth, doesn't this mean that *revolves around* relates the Sun to the Earth? If *revolve around* does relate the Sun to the Earth, is it not a *fact* that the Sun revolves around the Earth? Thus, the attempt to characterize a false proposition turns into the specification of something like a fact. But for Russell, a fact, as we saw, is a *true* proposition. Hence Moore–Russell propositions "leave[ ] the difference between truth and falsehood quite inexplicable" (*NTF*, 176).

1.3 Diagnosis and Multiple-Relation Theories

On the Moore–Russell conception, a proposition is a single unified entity constituted from the entities that it is about. In this conception, propositions play two roles:

- They are the objects of belief and other "propositional" attitudes; having a belief is standing in a relation to a proposition.
- If they are true, then they are facts or states of affairs.

This package of views runs into the problem of false propositions because of three intuitions:

(1) A fact obtains only if the constituents of that fact are connected with one another.
(2) Some of our beliefs are false.
(3) If a belief about certain entities standing in a relation is false, then it's *not* a fact that these entities are connected.

By (1), for propositions to play the fact role, they have to consist in connection of their constituents. By (2), in order for these propositions to play the objects-of-attitudes role, some of them have to be false. So false propositions have to consist of connections of their constituents. But false propositions are the

[6] Other reasons include (a) paradoxes stemming from assuming the existence of Moore–Russell propositions – see Stevens (2005) – and (b) consequences for acquaintance with such propositions of Russell's solution to the George IV puzzle of "On Denoting" (1905b); see Proops (2011).

objects of false beliefs. This contradicts (3), according to which if one has a false belief, then the entities the belief is about are *not* connected with one another. Put slightly differently, the problem is that a *false belief* turns out to be *just as much a unified complex* of entities in the world as a *true belief*.

In response, Russell in effect comes to the conclusion that the two roles have to be discharged by different things. He continues to hold that facts are connections of entities, which he now calls "complexes." He drops the position that beliefs and other attitudes consist of a relation to a single object consisting of a connection of entities.

If belief is not a relation to a single propositional object, what is it? Russell's answer is the so-called multiple-relation theory (MRT) of judgment. In the period that concerns us, between *The Principles of Mathematics* (*PoM*) and Wittgenstein's criticisms in 1913, Russell formulated some three versions of an MRT. The first and most basic was advanced in the introduction of *PM* (43–4) and stated in most detail in *NTF* (177–85).[7]

Belief is a *relation* of the *believer* to *multiple entities*. So, if

Ptolemy believes that the Sun revolves around the Earth,

then he stands in *a cognitive relation to three things*:

- the Sun,
- the Earth,
- the relation *revolve around.*

Ptolemy's having this belief *does not* involve the Sun standing in the *revolve around* relation to the Earth. Indeed, Ptolemy's belief is *false* because in reality, *revolve around* does *not* relate the Sun to the Earth; *there is no fact* of the Sun's revolving around the Earth. Falsity is the *nonexistence of a corresponding fact*.

If Elizabeth II believes that Churchill smoked, then she stands in the *believe* relation to Churchill and to the property of *smoking*. The Queen's belief is true because in reality, Churchill did have this property; he was in fact connected to *smoking*. *True belief* is the *existence of a corresponding fact*.

The theory remains anti-idealist because, in having a belief, *true or false*, a subject is in *direct contact with the objects* in the world her belief is about.

On this theory, a belief is a fact or state of affairs, a complex consisting of the *judge* relation uniting a judger and the entities the judgment is about. It is a state or episode of belief. In "Knowledge by Acquaintance and Knowledge by

[7] An account of belief with some elements of an MRT is sketched but not endorsed in Russell (1906b 46–7); here belief is a multiple relation holding of a believer and ideas of objects of belief, rather than the objects themselves.

Description" (Russell 1910, hereafter *KAKD*) and *Theory of Knowledge* (Russell 1913b, hereafter *TK*), this form of account is applied to supposition and understanding in addition to judgment or belief. For this reason, I call accounts of this form MRTs, but focus on belief. I will refer to the three versions of MRT we will examine by numerical subscripts on "MRT." Following Russell, I will call views of belief, supposition, and understanding as relations to a Moore–Russell proposition *dual-relation theories* (DRTs).

In order to discuss MRTs, let's fix some terminology drawn from *TK* (117), illustrated by Russell's most well-known example (taking *Othello* to be history rather than fiction):

(1) Othello believes that Desdemona loves Cassio

- The *subject* of this judgment is Othello.
- The *objects* of this judgment are Desdemona, *love*, and Cassio.
- The *terms* of this judgment are the subject and the objects: Othello, Desdemona, *love*, and Cassio.
- The *main relation* of this judgment is the relation *judge*.
- The *object-relation* of this judgment is the relation *love*.[8]
- The *constituents* of the judgment are its four terms together with the main relation.
- The *judgment* or *belief complex* is the complex described or expressed by (1), consisting of the main relation *judges* relating the terms. We also call such complexes *belief or judgment states* of the subject.

I specify a complex by mentioning its relating relation followed by its remaining terms, separated by commas and surrounded by double angle brackets. Complexes are unified entities constituted from their terms, not lists of terms, so commas merely enable mention of (occurrences of) terms, and do not indicate that the complex is a list.[9] So, the judgment complex described by (1) is

(2) *Judge*≪Othello, Desdemona, *love*, Cassio≫

Before moving on, I note an issue about MRT ascriptions of beliefs. When Russell held a DRT, he generally took statements as expressions of Moore–Russell propositions. Multiple-relation theories rule this out. In *PM*, Russell

[8] Although this is often called "subordinate relation" in the secondary literature, Russell himself only ever uses "subordinate verb," in Russell (1918, hereafter *PLA*, 59, 61) and Russell (1959, hereafter *MPD*, 88). I'm grateful to Rosalind Carey for this point.

[9] For reason to appear, I want to avoid prejudging whether the relating relation of a Russellian complex relates in some order, hence my avoidance of single angle brackets often used for ordered sequences, or braces often used for unordered collections.

takes a statement to be an "'incomplete' symbol; it does not have meaning in itself, but requires some supplementation in order to acquire a complete meaning" (*PM*, 44). He goes on to claim that "judgment in itself supplies a sufficient supplement" (*PM*, 44). But then, if ascriptions of judgments are statements, an infinite regress seems to threaten. It might then appear imperative to specify belief states by means other than statements. Perhaps for this reason, some commentators hold that the right way to specify beliefs is by definite descriptions.[10] Thus, I take expressions like (2) to be definite descriptions, which, if they denote, denote belief states.

Since MRTs are motivated by the failure of the Moore–Russell view of propositional entities to distinguish truth from falsity, they must include an account of truth and falsity. As noted, Russell's account is a correspondence theory of truth. The primary bearers of truth and falsity are judgment complexes. A judgment complex is true if a *corresponding complex*, which I will also call the *truth-making complex*, exists. The existence of the truth-making complex consists in the object-relation of the judgment relating the remaining objects of the judgment.[11] In our example, the corresponding complex exists if the object-relation *love* relates Desdemona and Cassio into a complex:

(3) *Love*≪Desdemona, Cassio≫

What makes Othello's belief state (2) false is that there is in the world no complex described by (3). Corresponding complexes are what Moore–Russell propositions, in their fact or state-of-affairs aspect, became in Russell's new theory.

On one philosophical conception, belief is a thinker's representation of the world. Russell's MRTs are consistent with such a conception; however, they differ in one way from a familiar idea of representation: they have no medium of representation. What I mean is this. We take, for example, a string of English words such as "Plato taught Aristotle" to represent a state of affairs, so the representation is composed of words, the words denote various objects, and putting words together in some specific way somehow represents how these objects are supposed to stand to one another in the world. The words are the medium of linguistic representation and in general they are distinct from the entities represented. For Russell, on the other hand, having a belief is just a matter of a thinker's standing in a relation to certain entities. So, states or episodes

[10] For example, Landini (1991, 38–9) and Levine (2013, 299n7). Of course, definite descriptions are also incomplete symbols, requiring to be supplemented, according to the theory of descriptions, to a statement.

[11] Note that talk of the existence and nonexistence of the truth-making complex is to be cashed out using the theory of descriptions.

of belief do not involve a thinker's standing in a relation to the constituents of a complex entity such as a sentence, in English or mentalese, which then represents worldly complexes. Nothing distinct from the connection of the believer with entities the belief is about is required for her to represent how these very entities are arranged in the world. This results again from anti-idealism: even if in believing truly we are *not* in direct contact *with facts*, nevertheless in having a belief, *true or false*, a subject is in *direct contact with the objects* in the world their belief is about; no mental or linguistic intermediaries intervene.

1.4 Problems of Multiple-Relation Theories

It is generally agreed that Russell's MRTs face two salient difficulties.

First, central to Russell's critique of Bradley's idealism are *nonsymmetrical relations*, relations *R* such that *a*'s standing in *R* to *b* does not imply *b*'s standing in *R* to *a*.[12] Given a dual – that is, dyadic – nonsymmetrical relation and two terms, there are, intuitively, two judgments that a subject can make about that relation and terms:

> Let us take the judgment "A loves B." This consists of a relation of the person judging to A and love and B, i.e. to the two terms A and B and the relation "love." But the judgment is not the same as the judgment "B loves A." (*NTF*, 183)

How does an MRT explain the difference, for example, between

(4) Othello believes that Cassio loves Desdemona

and

(1) Othello believes that Desdemona loves Cassio?

Othello might have had both beliefs, but they're different beliefs, since the belief described by (1) is false while that described by (4) is true,[13] and yet (1) and (4) mention the same objects of belief: *love*, Desdemona, and Cassio. More generally, the challenge for an MRT is to account for distinct judgment complexes composed of the same constituents, the same subject, and the same objects. Call this "the direction problem" (DirP).[14]

[12] Russell in fact focuses on a subset of nonsymmetrical relations, which he calls "asymmetrical," such that *a*'s standing in *R* to *b* implies *b*'s *not* standing in *R* to *a* (see, e.g., *PoM*, 218).

[13] I'm grateful to Mihaela Fistioc for pointing out that it is at best unclear from *Othello* that Cassio loves Desdemona, so taking (4) to be true involves a double de-fictionalization.

[14] This name comes from Griffin (1985, 219).

The individuation of beliefs such as (1) and (4) with the same objects is only one aspect of DirP. We get at two others by noting again that a principal reason why we take these beliefs to be distinct is that one is true and the other false. A belief is true just in case its objects constitute a truth-making complex, and these beliefs have the same objects. So, in order for them to have distinct truth-values, it seems that there must be something like two ways in which the objects can be related, such that objects being related in one way makes one of these beliefs but not the other true, and vice versa if the objects are related in the other way. The two "ways" in which a single set of objects of belief can be related are distinct truth-conditions for beliefs about these objects. So, the problem of explaining how distinct beliefs about the same objects can have distinct truth-values factors into two. One is to specify distinct truth-conditions for beliefs about the same objects. The other is to specify which of these truth-conditions makes which of the individuated beliefs true. The direction problem has three components:

1. The problem of belief individuation (Prob_{Bel}): how to distinguish between intuitively different beliefs with the same objects.
2. The problem of truth-conditions individuation (Prob_{TC}): how to distinguish between truth-conditions of such beliefs having the same objects.
3. The problem of coordination between beliefs and truth-conditions ($\text{Prob}_{\text{Coord}}$): how to specify which of the individuated truth-conditions makes which of the individuated beliefs true.

The second salient difficulty for the MRTs is to distinguish judgments such as (1) and (4) from whatever might be expressed by:

(5) Othello believes that Cassio Desdemona loves

There are two issues here. First, does this sentence describe a belief at all? Is the sequence of words

(6) Cassio Desdemona loves

a grammatically well-formed sentence of English? If not, is it meaningful? (We will revisit these questions in Section 2.5.1.) If (6) isn't meaningful, then (5) seems to describe Othello as judging nonsense, but is there really such a thing as judging nonsense?

Second, even if it is possible to judge nonsense, how is this purported judgment distinguished from (1) and (4)? Again, all of these statements mention the same three objects of belief. Let's call these issues collectively "the nonsense problem" (NonP).[15]

The nonsense problem does not arise for the old DRT because of a feature of Moore–Russell propositions Russell adopts in *PoM*: in each proposition, a *relating relation* unifies the terms into that proposition, and only relations can do this job. Perhaps the best we can do to interpret the apparently nonsensical sentence (6) is to take it as seeming to express a proposition in which Desdemona is the relating relation. But Desdemona and other human beings are what in *PoM* Russell calls *things*, and things do not play this relating role. Hence these sentences express no propositions.

1.5 Multiple-Relation Theories, Versions 1 and 2

In *NTF*, Russell provides a solution to DirP on the basis of a doctrine that he held since *PoM*: nonsymmetrical relations such as *love* have *sense* or *direction*:

> We may distinguish two "senses" of a relation according as it goes from A to B or from B to A. Then the relation as it enters into the judgment must have a "sense"; and in the corresponding complex it must have the same "sense." Thus the judgment that two terms have a certain relation R is a relation of the mind to the two terms and the relation R with the appropriate sense: the "corresponding" complex consists of the two terms related by the relation R with the same sense. (*NTF*, 183–4)

Russell also phrases this account by the claim that in judging "the relation [R] must not be abstractly before the mind, but must be before it as proceeding from A to B rather than from B to A" (*NTF*, 183).

Two complexes may be constituted by a dual object-relation's relating the other two objects of a belief in distinct senses. Existences of these complexes are distinct truth-conditions that solve Prob_{TC}. The senses of the object-relation "as it enters the judgment," or "before the mind" of the subject individuate belief states such as (1) and (4), thereby solving Prob_{Bel}. The existence of the complex in which the object-relation relates in the same direction as it has in the belief is the truth-condition of that belief; this solves $\text{Prob}_{\text{Coord}}$.

Call an MRT that includes this solution to DirP MRT_1.

It's not altogether clear, however, how this DirP solution is supposed to work. The sense of *love* clearly allows for distinguishing two *complexes*

(7) *Love* < Desdemona, Cassio >

[15] Griffin calls this the "wide form of the direction problem" (1985, 219), which sounds odd, since it's not clear what this issue has to do with direction. However, we will see that there is an underlying unity to these problems.

(8) *Love* < Cassio, Desdemona >

where I use single angle brackets to highlight the order of relating. On the old DRT, these complexes would be Moore–Russell propositions, and Othello's being related by *judge* to them accounts for the distinction between (1) and (4). Since Russell's account is supposed to be an MRT, his characterizations – "a relation of the mind to the two terms and the relation R with the appropriate sense" and "having the relation R before the mind as proceeding from A to B" – can't mean the mind's being related by *believe* to one or the other of (7) and (8). Indeed, there is in reality no complex corresponding to (8). So, what do these characterizations in Russell's account mean?

This question underlies a criticism of MRT_1 made by Stout, who sees "the relation of loving [being] apprehended as proceeding from A to B" as conceding that the "items to which the mind is related in judging do have a unity of their own and are apprehended as having a unity of their own" (1910 202–3). Stout in effect claims that invoking sense of relations to solve DirP makes MRT_1 collapse into the earlier DRT.

Here's another formulation of Stout's point. According to the basic MRT, the object-relation *R* is just one of the terms unified into belief complexes by the subject's judging. But then it doesn't occur as a relating relation in the judgment complex. So, the object-relation doesn't unify the other objects of the judgment complex. *Love* doesn't unify Desdemona to Cassio in Othello's judgment (1), even if it has an intrinsic direction when it relates, so in what sense does it "enter" that judgment as "proceeding" "from" Desdemona "to" Cassio?[16]

Russell responds to Stout's criticism with what he, in a letter to Stout, calls a "a slight re-wording of the account of sense in judgment" (quoted in Stout 1910, 203). This "re-wording" is incorporated in the next version, MRT_2, in *The Problems of Philosophy*:[17]

> [T]he relation of judging has what is called a "sense" or "direction." We may say, metaphorically, that it puts its objects in a certain *order*, which we may indicate by means of the order of the words in the sentence. (In an inflected language, the same thing will be indicated by inflections, e.g. by the difference between nominative and accusative.) Othello's judgement that Cassio loves Desdemona differs from his judgement that Desdemona loves Cassio, in spite of the fact that it consists of the same constituents, because the relation of judging places the constituents in a different order in the two cases. (Russell 1912a, hereafter *PoP*, 198)

[16] For further discussion of Stout's arguments, see Lebens (2017) and van der Schaar (2013).

[17] It may be said that the MRT here is what Russell had in mind already in *NTF*, since he calls it a "re-wording"; perhaps one should label it $MRT_{1.5}$.

This DirP solution changes two components of the MRT_1 solution. The problem of belief individuation is now solved by distinct directions in which the *main* relation, *believe*, relates, rather than directions that the object-relation somehow "has in" the beliefs. The problem of coordination between beliefs and truth-conditions, accordingly, is answered by matching the order of object-relating in truth-making complexes to the order of *believe*'s relating.

There's some uncertainty about how this answer is supposed to work. Russell doesn't specify what exactly are the orders in which *believe* relates to form belief complexes. All he says is that the order of believing is "indicated" by "the order of the words" in judgment ascription sentences, or by the cases of words in belief ascription sentences of inflected languages. How is this "indication" supposed to work? Does (1) describe the complex

(9) *Believe* < Othello, Desdemona, *love*, Cassio >

because the order of believing in (9) matches the *left-to-right order* of the words in (1) that denote the terms of (9)? But in complex

(10) *Believe* < Cassio, *love*, Desdemona, Othello >

the order of believing matches the *right-to-left order* of words in (1), so why doesn't (1) describe (10)?

Russell's characterization of truth-making complexes in *PoP* helps somewhat:

> [I]f Othello believes *truly* that Desdemona loves Cassio, then there is a complex unity, "Desdemona's love for Cassio," which is composed exclusively of the objects of the belief, in the *same order as they had in the belief*, with the relation which was one of the objects occurring now as the cement that binds together the other objects of the belief. (*PoP*, 200; emphases mine)

So, whatever the judgment complex described by (1) might be, its truth-making complex is

(7) *Love* < Desdemona, Cassio >;

moreover, in that judgment complex, *believe* relates Cassio and Desdemona in the "same order as" *love* relates them in (7). This suggests that the exact order in which *believe* relates *love*, Othello, Desdemona, and Cassio doesn't matter. What matters is whether Desdemona comes "before" or "after" Cassio in the order of believing. Thus, perhaps Russell has in mind that any of the following orders of believing

$$Believe < \text{Othello}, love, \text{Desdemona}, \text{Cassio} >$$
$$Believe < \text{Desdemona}, \text{Othello}, love, \text{Cassio} >$$
$$Believe < love, \text{Desdemona}, \text{Cassio}, \text{Othello} > ,$$

along with (9), may be the belief complex described by (1), because in all four of these complexes, Cassio "is after" Desdemona in the order of believing, matching their order in (7).

This suggests that for Russell, the order in which judging relates is not linear but partial. Let's write

$$x \rightarrow_{R,C} y$$

for

In the complex C formed by the relating of R,

R relates (immediately) "from" x "to" y.

Then, perhaps, where R is the relation of judging, J, the subject S is least ("first") in the ordering, the object-relation O "follows" S

$$S \rightarrow_{J,C} O,$$

the remaining objects of belief are linearly ordered by $\rightarrow_{J,C}$,[18] and, maybe the $\rightarrow_{J,C}$-least object of judgment x "follows" the subject

$$S \rightarrow_{J,C} x$$

but no object either "precedes" or "follows" O. (I omit "C" in general from now on.) Here's a not altogether satisfactory notation for this idea:

$$Believe < \text{Othello}, love; < \text{Desdemona}, \text{Cassio} >>$$

with the semicolon indicating "neither precedes nor follows." Note that I'll continue to use single angle brackets and commas for all other ordered relating.

2 Wittgenstein against Multiple-Relation Theories

In the summer of 1913, Russell worked on a manuscript on the theory of knowledge, now published as *TK*, and in it yet a third version of the MRT appears. We have Russell's own testimony, in letters to Lady Ottoline Morell, that during the composition of this manuscript, Wittgenstein criticized it, in particular Russell's views on judgment. Russell also wrote Lady Ottoline of making attempts to "circumvent" Wittgenstein's attacks. In a letter to Russell,

[18] Nothing stands in the way of taking $\rightarrow_{J,C}$, restricted to the objects of belief, to have an arbitrary order type. Gödel seems to have considered infinitary MRTs; see Floyd and Kanamori (2016).

Wittgenstein writes, "I am very sorry to hear that my objection to your theory of judgment paralyses you" (July 22, 1913 Wittgenstein 2008, hereafter *WC*, 42). Russell abandons the unfinished manuscript. All this suggests that Wittgenstein's criticisms might have played some role in Russell's formulation or possibly also his rejection of MRT_3. The correspondence unfortunately provides little explanation of the criticisms or the responses. So, any attempt to specify what exactly Wittgenstein's criticisms are, and what their relation is to MRT_3, has to be based on the meager clues of the correspondence, the chronology of Wittgenstein's interactions with Russell, and *TK*. No such attempt can avoid a large measure of conjecture, and the account I will be developing is no exception. I claim for it philosophical interest first, historical accuracy second.

2.1 MRT_3

I start with some relatively uncontroversial points about MRT_3, the chronology of Wittgenstein's discussions with Russell during the writing of *TK*, and Wittgenstein's formulations of the problems afflicting the MRTs. I will refer to Chapter y of Part x of *TK* as $x.y$.

2.1.1 Main Differences between MRT_3 and MRT_2

There are three significant changes from MRT_2. First, in *TK*, Russell argues that certain nonsymmetrical temporal relations, expressed by "before" and "after" do *not* have sense, do *not* relate in a specific order. Russell's argument was rediscovered by Fine (2000), who calls relations without sense "neutral"; I'll refer to Russell's conclusion as *neutralism* about temporal relations, and the view it opposes *directionalism*. In MRT_3, Russell does not invoke the sense of either the main or the object-relations.

Second, in MRT_3, each judgment complex has an additional constituent, a *form*.

Third, in MRT_3, beliefs Russell calls "permutative" (PBs) are characterized in terms of a notion of *positions* of constituents in a complex. Russell initially uses positions to account for features of nonsymmetrical temporal relations without relying on their having sense; this account was also rediscovered by Fine (2000), who calls it *positionalism*.

Not clear are the three questions on which we will focus:

- Why did Russell move from MRT_2 to MRT_3?
- Why did he introduce forms?
- Why did he introduce a theory of positions, and why did he use it in a theory of permutative beliefs (TPB)?

2.1.2 Chronology

All dates are in 1913, with the month first.

5/20	Wittgenstein goes to Russell with a "refutation of a theory [Russell] used to hold." This happens just before Russell begins I.vii. Call this $Crit_1$. Russell writes that Wittgenstein "was right, but I think the correction required is not very serious."
5/21–2	Russell begins and finishes I.vii. In I.vii, forms make their first appearance, and Russell first argues for neutralism about temporal relations and introduces positionalism about them.
5/23	Russell finishes I.viii–ix.
5/24	Russell begins Part II.
5/25	Russell works on understanding in II.i.
5/26	Wittgenstein presents Russell with a second round of "inarticulate" criticisms in response to "a crucial part of what [Russell has] been writing," criticisms Russell feels "in [his] bones … must be right." Call this second criticism $Crit_2$.
5/30	Russell reaches the end of II.iv.
5/31	Russell writes to Lady Ottoline that he could "circumvent Wittgenstein's problems."
6/1	Russell finishes the treatment of truth, manuscript page 300. This might be the end of II.v.
6/6	Russell reaches page 350, the last of the extant manuscript.
Before 6/18	Wittgenstein formulates an "exact" expression of his criticism.
7/22	Wittgenstein writes that he's sorry his criticisms "paralysed" Russell.

2.1.3 Wittgenstein's Criticisms

Wittgenstein formulates some six (versions of) objections to Russell's MRTs:

1. I can now express my objection to your theory of judgement exactly: I believe it is obvious that, from the proposition "A judges that (say) *a* is in a relation *R* to *b*," if correctly analysed, the proposition "$aRb. \vee .{\sim}aRb$" must follow directly without the use of any other premiss. This condition is not fulfilled by your theory. (Letter to Russell, 6/1913, *WC*, 40)
2. When we say A judges that etc., then we have to mention a whole proposition which A judges. It will not do either to mention only its constituents, or its constituents and form, but not in the proper order. This shows that a proposition itself must occur in the statement that it is judged; however, for instance, "not-*p*" may be explained, the question what is negated must have

a meaning. (Wittgenstein 1913, hereafter *NL*, Summary; this is a set of notes dictated to a stenographer in Birmingham and to Russell in Cambridge, now called "Notes on Logic")

3. Every right theory of judgement must make it impossible for me to judge that this table penholders the book. Russell's theory does not satisfy this requirement. (*NL*, third MS)
4. A proper theory of judgement must make it impossible to judge nonsense. (*NL*, 95)
5. There is no thing which is the form of a proposition, and no name which is the name of a form. Accordingly we can also not say that a relation which in certain cases holds between things holds sometimes between forms and things. This goes against Russell's theory of judgement. (*NL*, fourth MS 105[4])
6. The correct explanation of the form of the proposition "A judges *p*" must show that it is impossible to judge a nonsense. (Russell's theory does not satisfy this condition.) (5.5422)

2.2 A Compass for the Perplexed

There is a vast and growing literature on Wittgenstein's philosophical interactions with Russell in summer 1913. To provide readers with some initial bearings, I outline principles of classification based on three questions:[19]

1. Are Wittgenstein's criticisms focused on NonP or DirP?
2. How are Wittgenstein's criticisms connected to Russell's abandonment of direction of (temporal) relating, and to Russell's account of PBs in terms of positions in complexes?
3. How effective are Wittgenstein's criticisms against the MRTs?

2.2.1 Nonsensism, (Non)Conformism, Typism, Unitarianism

Most commentators plump for NonP as the focus of Wittgenstein's attack. Call this choice *nonsensism*. Nonsensism seems all but mandatory, given criticisms 3, 4, and 6. Criticisms 4 and 6 state that an adequate theory of judgment must rule out judgments of nonsense, and criticism 6 explicitly charges Russell's theory with failure to meet this adequacy condition. According to criticism 3, Russell's theory fails to rule out judging that "this table penholders the book," which doesn't seem to be a meaningful sentence. Criticism 1 may also be read

[19] See Connelly (2014) and Lebens (2017) for detailed presentations and analyses of some of the interpretations sketched later in this Element.

as supporting nonsensism, for the instance of the law of excluded middle Wittgenstein cites, "*aRb*.∨.~*aRb*," may be understood to mean or imply the claim that "*aRb*" has a truth-value, which nonsense does not.

Most nonsensists hold that the introduction of logical forms in MRT_3 is intended to solve NonP, and that $Crit_2$ is intended to show that they can't. Many of these nonsensists do *not* take $Crit_2$ to focus on the coherence of Russell's notion of forms; call these *conformists*.

Two branches of conformist nonsensism are followers of the influential interpretation of Sommerville (1981) and Griffin (1985), and those who reject this "standard" interpretation. The standard reading characterizes the problem that Wittgenstein sees in MRT_3 as follows. In order to solve NonP, Russell has to supplement MRT_3 with restrictions from the theory of types of *PM*, restrictions that Sommerville and Griffin take to govern what kinds of entities may occur in a judgment complex. But Russell also uses the MRT to justify the theory of types. Hence, Russell cannot use type restrictions to shore up the MRT without circularity. Call this reading *typism*.

The typist reading has been criticized on several grounds. Landini (1991, 2007) and Stevens (2003) argue that the MRT doesn't justify the type hierarchy, but rather the order hierarchy of *PM*, so that there is no circularity as alleged by the typist reading. Hanks (2007) argues that Russellian restrictions are able to resolve NonP. Potter (2009, 130) holds that the type-theoretic argument is too "elaborate" for Wittgenstein, whose arguments "are simple."

Except Landini, these critics hold that the problem Wittgenstein discerns in the MRT is its failure to solve the problem of the unity of the proposition. This is usually understood as based on an intuitive distinction between a list or set of entities and a statement or proposition in which the entities are unified. Call this style of interpretation *unitarianism*.

Not all unitarians are nonsensists *stricto sensu*. Hanks (2007), for instance, holds that Wittgenstein's principal insistence is that judgment involves a unified propositional entity having a truth-value, something missing in the MRTs.

Typism hasn't vanished; Connelly (2021, 2011, 2014) advocates a *reformed typism*, in which types are metaphysical distinctions rather than the logical types of *PM*, and $Crit_2$ is the complaint that Russell has to supplement MRT_3 with metaphysical restrictions to attain a viable account of judgment.

A final branch of nonsensism consists of commentators who take $Crit_2$ to be directed at the coherence of Russell's conception of logical forms. The most prominent of these *nonconformists* is Pears (1977, 1989). See also Carey (2007).

Most nonsensists don't take Russell's neutralism about temporal relations or his positionalist account of PBs to have much to do with Wittgenstein's criticisms.

2.2.2 Ordinalism: Wittgensteinians versus Russellians

In spite of the apparently overwhelming textual grounds in favor of nonsensism, DirP isn't without its champions. Call them *ordinalists*. Ricketts (1996), for instance, has urged that Wittgenstein's criticisms led Russell to hold that the relation *believe* doesn't have sense, which then undermines MRT_2's solution to DirP. Pincock (2008) has extended this line of interpretation to discern a "devastating" difficulty in Russell's TPB. MacBride (2013) agrees mostly with Pincock's analysis of the problems with TPB, but holds that the MRT is easily rescued from them simply by abandoning neutralism about relations.

Ricketts and MacBride are exemplars of opposite answers to question 3. MacBride holds that although Wittgenstein managed to land on a problem for Russell's MRTs, it is easily dismissed. Scholars who, like MacBride, take Wittgenstein's criticisms to be largely ineffective against the MRT include Lebens and Landini. Call *Wittgensteinians* them *Russellians*. Call those who, like Ricketts and Pincock, hold that Wittgenstein had located deep difficulties for the MRT. Not all Wittgensteinians are ordinalists; some, such as Hanks (2007) and Johnston (2012), are unitarians; others are the original typists Sommerville and Griffin.

2.2.3 Contemporary Multiple-Relation Theories and Magicians

Multiple-relation theories have staged something of a revival in recent analytic philosophy, starting roughly with Jubien (2001), who assumes that "human beings [are] sources of intentionality" that "represent things that are entirely independent ourselves" (55).

On Russell's MRT, initially anyway, there's no more to judgment than standing in a relation to a bunch of objects. Contemporary versions, following Jubien, delve into or perhaps rename that relation as some special representational feature or capacity of (human) minds. Jubien abjures any explanation of how this representation "happens" (55). Others hold that the special capacity is for predicating a property or relation of object(s).

Indeed, some commentators hold that the real difficulties with the MRTs are resolvable by taking (human) minds to have this special feature or capacity. Call them *magicians*. Soames (2014) announces that no one before him had understood the *real* problem of the unity of the proposition, which is to explain how any structure of entities could represent the world and thus be true or false, and states that Russell fails to solve *this* problem. Later, Soames grudgingly allots to Russell and Wittgenstein some insight into this problem before declaring that only he has in fact discovered the solution, namely the structures in question have to be employed in cognitive acts of predication, which are apparently supposed to possess *virtus repraesentativa* (Soames 2017). Hanks (2015) espouses a closely connected view of predication as the key ingredient of propositional unity, although it's not clear whether he connects the idea to Russell's MRTs. Lebens (2017) is a magician who agrees with Soames that explaining representation is the principal problem for the MRTs, but holds that Russell himself understood this and that a Russellian MRT can solve it by recourse to predication. A different strain of magician is Landini, who is closer to Jubien in holding the preferred vehicle of *virtus repraesentativa* to be a version of Brentano's (1874) notion of intentionality, and who takes Russell to have adopted Brentanian intentionality to account for how minds home in on truth-making complexes.

It may surprise the reader that I count Hanks–Soames theories as MRT variants, since they are theories of propositions. But in fact Russell himself suggests that Moore–Russell propositions may be (logically re)constructed as "something which a number of mental events have in common" (see *TK*, 114; see Section 2.4.5), and such events include "act[s] of belief" (*PoP*, 197). This is not to say that all contemporary MRT-inspired accounts are in accord with Russell's stance in this period; Moltmann (2003), for instance, takes propositions to be types of the products of predicational and other cognitive acts.

2.2.4 Quasi-Aufhebung

I now outline the view of Wittgenstein's criticisms of the MRTs I will propose. From Wittgenstein's perspective, NonP and DirP are aspects of a more general difficulty for the MRTs, that of providing determinate truth-conditions for judgment complexes. For this reason, I'm neither Haredi nonsensist nor Haredi ordinalist. In Section 2.3, I show that many of the problems for MRTs proposed by nonsensists can be solved provided that Russell can rely on directionalism about believing and understanding. I'm agnostic about whether Wittgenstein talked Russell into neutralism about these cognitive relations. However, in Section 2.4, I show that Russell's positionalist TPB fails to solve DirP, for reasons similar to those advanced by ordinalists, a result that also blocks a solution to NonP. I speculate that one part of $Crit_2$ may be to urge this result on Russell. In Section 2.5, I argue that rejecting neutralism about belief doesn't, in the end, help. Specifically, there is no coherent account of the sense of believing consistent with MRTs, and this problem may be another target of $Crit_2$.

On my reading, the two problems presented by $Crit_2$ share two features:

- Both are MRTs' failure to explain how beliefs have determinate truth-values.
- In both cases, positing complexes constituted from objects of belief overcomes this failure.

The conundrum is that such complexes seem to be none other than Moore–Russell propositions with their attendant problem of falsity. Support for my speculation comes from Russell's reaction to $Crit_2$ and Wittgenstein's initial attempt to resolve Russell's difficulties through a theory of propositions in *NL*; both aim to find a way out of this conundrum.

2.3 Nonsensism

Both of the innovations of MRT_3 first appear in I.vii, which Russell begins to write the day after $Crit_1$. It's natural to speculate that one or the other of these innovations resulted from $Crit_1$. In this section, I develop one such line of speculation. In brief, the proposal is that $Crit_1$ led Russell to posit forms as a response to NonP, and $Crit_2$ showed Russell the inadequacy of this response.

2.3.1 The Role of Forms

Russell took $Crit_1$ to apply to a "theory he used to hold," and thought that the "correction required" to address $Crit_1$ "is not very serious." Russell writes that "we must understand the 'form' before we can understand the proposition," where by "proposition" he means statement, and continues, "I held formerly that the objects alone sufficed, and that the 'sense' of the relation of understanding would put them in the right order; this, however, no longer seems to me to be the case" (*TK*, 116). This suggests that the theory Russell used to hold is MRT_2, on which relations such as *understand* relate a subject and the objects of his thought in some order, and that the correction Russell took to be required in response to $Crit_1$ is the addition of forms. Perhaps Russell estimates this correction as "not very serious" because he had already been considering logical forms late in 1912 (see Russell 1912b, 55–6; for further discussion, see Pears 1989; Carey 2007).

But why is the order in which *understand* relates not enough?

To understand Russell's thinking, let's first see what role form is supposed to play in understanding:

> [I]n order to understand "*A* and *B* are similar," we must know what is supposed to be done with *A* and *B* and similarity, i.e. what it is for two terms to

> have a relation; that is, we must understand *the form of the complex which must exist if the proposition is true*. (*TK*, 116; emphases mine)

I take "what is supposed to be done with" the objects of understanding to be the way in which these objects are supposed to constitute the truth-making complex. This "way" is the form of the truth-making complex. In this example, the truth-making complex is constituted by "two terms having a relation."

Now, suppose someone is forming a shopping list in their mind, or listing alphabetically the people who went to their class reunion. Either of these states of mind, it would seem, may plausibly be taken to consists of a subject standing in some relation to items in some order. Neither, however, seems clearly to require the subject to have a conception of the items as constituting a fact. One shortcoming of MRT_2 is that mere order in understanding or believing is not enough to distinguish these states from thinking of a list.

It's still not clear, however, whether remedying this shortcoming with the addition of forms provides any answer to NonP.

To make progress, let's start with Russell's characterization of the "natural symbolic *expression* for the form of a given complex" as

> the expression obtained by replacing the names of the constituents of the complex by letters representing variables, using different kinds of letters for constituents of different logical kinds .... Thus we may indicate the general form of a dual complex by "*xRy*." (*TK*, 113)

So forms are like templates or schemata, showing how a set of entities assembles into a complex. More important, this "how" is spelt out in terms of "logical kinds." These "logical kinds" are metaphysical categories of entities determined by how they "enter into" complexes; I will call these ontological categories.[20] Two broad categories are particulars and universals:

> A particular is defined as an entity which can only enter into complexes as the subject of a predicate or as one of the terms of a relation, never itself as a predicate or relation. (*TK*, 55–6)

"Entering" a complex "as subject," as opposed to entering "as predicate or relation," are different *modes of occurrence* of an entity in a complex, or different *roles* an entity plays in a complex. This idea of modes of occurrence or

[20] I follow Pincock (2008) in holding Russell's "logical kinds" not to be the type distinctions of the theory of types of *PM*, nor based on type-theoretic distinctions. Linsky (1999) is one of the first to distinguish Russell's metaphysical categories from type-theoretic "logical categories" and my inspiration for the label "ontological." Ontological categories are akin to what Landini (2007, 2014) calls "type* distinctions."

roles goes back to the *PoM* doctrine that in every Moore–Russell proposition there is one constituent, a predicate or relation, that does the job of unifying the constituents into that proposition. Things, entities other than predicates and relations, never play this unifying role. In *TK*, this becomes the general ontological principle that only universals, never particulars, occur in relating modes or roles in complexes.

Understanding, for example, the statement "*A* and *B* are similar," according to Russell, involves "bringing" *A*, *B*, and the object relation *similarity*

> into relation with the general form of dual complexes. The form being "something and something have a certain relation," our understanding of the proposition might be expressed in the words "something, namely *A*, and something, namely *B*, have a certain relation, namely similarity." (*TK*, 116)

Thus, given their acquaintance with a form, the subject who understands a statement correlates the object-relation of their understanding with the relating-relation variable of that form, and the remaining objects with the remaining variables of the form, which is to understand that statement as true if a (unique) complex constituted by the object-relation relating the remaining objects exists.

Note that, if Russell accepts directionalism, as in MRT_2, then he may hold, in addition, that

- the non-relating-relation variables of every form occur in a linear order, and
- the correlation of the non-object-relation objects of belief with the non-relating-relation variables of the form of understanding preserves the linear order of the objects.

This upholds the MRT_2 solution to DirP by guaranteeing a unique corresponding complex.

We now connect forms to NonP. We are to explain why, for example,

(5) Othello believes that Cassio Desdemona loves

seems to describe a judgment of nonsense, while

(1) Othello believes that Desdemona loves Cassio

evidently describes a false belief. How could this be, if these statements refer to the same five entities? Here is a Russellian answer based on forms. The complement clauses of the (purported) belief ascriptions (5) and (1), respectively

(6) Cassio Desdemona loves

and

(11) Desdemona loves Cassio,

don't *merely* pick out the objects of the belief states. They *also* present the form of the truth-making complexes and, more important, how the objects of the belief constitute those complexes. In particular, following Russell's phraseology in the passage from *TK* quoted earlier in this Element, (11) presents the truth-making complex of belief state (1) as

(12) Something, namely Desdemona, and something, namely Cassio, have a certain relation, namely *love*.

That is, (11) presents the truth-making complex as having the form of dual complexes, and as constituted by *love* occurring in the relating role together with Cassio and Desdemona. The complement clause (6), in contrast, presents the truth-making complex of belief state (5) as

(13) Something, namely Cassio, and something, namely *love*, have a certain relation, namely Desdemona.

This presents the truth-making complex of (5) also as having the form of dual complexes, but as constituted by Desdemona occurring in the relating role together with Cassio and *love*. Now, as we saw, for Russell, it is a general (ontological) truth that no objects occur in the relating role in a complex. The existence of the truth-making complex of (5) – that is, a complex described by (13) – is logically incompatible with this general principle. In contrast, although there is in reality no truth-making complex for (1) as described by (12), the existence of this complex is not incompatible with the general principle. This difference provides a Russellian reconstruction of our intuitive sense that while (1) unproblematically describes a belief, (5) fails to do so because (6) is nonsense.

The order of believing by itself, without the correlation of objects to form, tells us nothing about whether any object of belief ever plays the relating role in any complex. Hence the foregoing response to NonP is not available from MRT_2.

So we can now speculate that Wittgenstein's $Crit_1$ is that MRT_2 fails to solve NonP, and that Russell's response to $Crit_1$ is to introduce forms, on the day after $Crit_1$.

2.3.2 Crit$_2$ and the Nonsense Problem

Given the reading just sketched of Crit$_1$, Crit$_2$ would naturally be taken to be some argument for the conclusion that positing forms fails to solve NonP after all.

One proposal begins from a letter to Russell dated January 1913 in which Wittgenstein considers an analysis of propositions involving forms.

> [I]f I analyse the prop[osition] Socrates is mortal into Socrates, Mortality and $(\exists x,y)\varepsilon_1(x,y)$ I want a theory of types to tell me that "Mortality is Socrates" is nonsensical, because if I treat "Mortality" as a proper name (as I did) there is nothing to prevent me to make [*sic*] the substitution the wrong way round. (*WC*, 38)

"$(\exists x,y)\varepsilon_1(x,y)$" is an expression of form, and Wittgenstein appears to be worried about the way in which "switching the order" of substitution of expressions into a form results in something "nonsensical." So something about "substitutions" of objects in a form shows that forms by themselves fail to solve NonP, and that a theory of types is further required.

The proposal goes as follows.[21] Russell holds a principle of substitution: if an entity occurs as term rather than relating relation in some complex, it can be substituted for another entity that occurs as term in another complex to yield a logically possible complex. For example, Socrates occurs as term in the dual complex described by

(14) Socrates preceded Plato

and Aristotle occurs as term in the complex described by

(15) Plato is older than Aristotle

So we can substitute Socrates for Aristotle in (15), and the result will be a possible complex:

Plato is older than Socrates

But the relation *older* also occurs *as term* in the complex

Older is a dual relation

So we can substitute *older* for Socrates in (14) to obtain a possible complex described by

[21] Considerations about substitution are regularly mentioned by conformist nonsensists, but Johnston (2012) provides a detailed formulation, of which the following is a simplified version.

(16) *Older* precedes Plato

Applying the substitution principle to judgment complexes leads to trouble. All objects of a belief occur *as terms* in the belief state. This includes the object-relation, even though it would occur as relating in the truth-making complex. Thus, in

(17) *Judge* < Othello, *love*; <Desdemona, Cassio>; *ξ* < *x*,*y* >>

where "*ξ* < *x*,*y* >" indicates the form of dual complexes, Cassio, Desdemona, and *love* all occur as terms, even though if its truth-making complex

(7) *Love* < Desdemona, Cassio >

exists, *love* occurs therein as relating relation. The substitution principle then implies that the result of swapping any of these objects of belief for another in complex (17) is a logically possible complex. Hence the following result of permuting the objects of (17),

(18) *Judge* < Othello, Cassio; <Desdemona, *love*>; *ξ* < *x*,*y* >>,

is a logically possible judgment complex.

Now, if (17) is the complex described by

(4) Othello believes that Desdemona loves Cassio,

then surely (18) is the judgment complex described by

(5) Othello believes that Cassio Desdemona loves,

which is apparently the ascription to Othello of belief in a piece of nonsense. So MRT$_2$, even augmented with logical forms, fails to solve NonP. Let's call the foregoing line of thinking the *substitution argument*.

2.3.3 Troubles with the Substitution Argument

By itself the substitution argument doesn't work against Russell.

We first note that the freewheeling use of "logically possible" in the substitution argument stands in some tension with Russell's rejection of modality as philosophically fundamental. Russell holds that there is no substantive notion of possibility, logical or otherwise: no proposition is possibly true or false, as opposed to plain true or false; nothing is a merely possible existent, as opposed to existing; nothing has merely possible being, and all terms subsist. Talk of the "logically possible" would need accounting for in more fundamental terms.

Next, note that Moore–Russell DRTs are not vulnerable to the substitution argument. In *PoM*, Russell holds that the result of substituting a term for a constituent of a Moore–Russell proposition is a *subsistent complex*, a complex that has being, even if not spatiotemporal existence.[22] But not all such substitutions result in a *proposition*. In the proposition

(7) *Love* < Desdemona, Cassio >

love occurs as a relating relation, and Desdemona occurs as term. So, if we swapped *love* and Desdemona, in the result, *love* occurs as term, but so does Desdemona, since Desdemona is a thing and things never occur as relating or predicating. So the result of the mutual substitution is a collection or an aggregate rather than a proposition.[23] But judgment and cognitive relations like it hold only between subjects and propositions. So, even if we take it that aggregates or collections count as nonsense, there is no judgment of nonsense.

Now I show that, in the context of an MRT, Russell can give an account of the *phrase* "logically possible complex" on which he is not committed to the subsistence of such complexes.

In *PM* (47) and *NTF* (175), Russell combines the rejection of propositional entities with the view that "phrases" purporting to express such entities are incomplete symbols. So, although his considered position is that there are no merely logically possible *complexes*, it is open to him to provide an account of *incomplete symbols* purporting to express such complexes. Here is a proposal deriving from *TK* (111). The *statement*

(19) The complex described by

Judge < Othello, Cassio; <Desdemona, *love*>; ξ < x,y >>

is logically possible

is true if and only if there exists a complex whose constituents are of the same ontological categories as those denoted by the names in the "phrase":

Judge < Othello, Cassio; <Desdemona, *love*>; ξ < x,y >>.

And statement (19) *is* indeed true, because

[22] What exactly does this talk of "substitution" really mean? Who or what does this "substituting"? The cash value of "substitution" in *PoM* consists in a principle of plenitude of complexes: all complexes composed of all terms subsist, although only some exist. Of course this principle teeters on the precipice of paradox. For more discussion, see Shieh (2019, chapter 8).

[23] For discussion of the distinctions among propositions, collections, and aggregates in *PoM*, see Shieh (2019, chapter 8).

(17) $\mathit{Judge} < \text{Othello}, \mathit{love}; <\text{Desdemona, Cassio}>; \xi < x,y >>$

describes an existing complex.

More generally, the *statement*

The complex Γ is logically possible,

where Γ is an *expression* specifying the constituents of a purported complex, is true just in case there exists a complex γ such that for every ontological category κ, the number of entities of category κ in the set of constituents of γ is the same as the number of entities of category κ in the set of entities denoted by names in Γ.

This shows that Russell is not committed to there being any nonsense judgments. However, it might be said that this isn't good enough, since it is consistent with all we have shown, and so, with all that we know, that there are beliefs described by (18). It might be said that Wittgenstein would demand more; he would demand that a theory of judgment rule out nonsense judgments rather than merely fail to imply their existence.

Russell can meet this demand in at least two ways. We have already seen, at the end of Section 2.3.1, the idea that purported nonsense judgments complexes are those the existence of whose truth-making complexes is logically incompatible with an ontological generalization. But it might be objected that on this proposal, such complexes as (18) are false beliefs since there are no complexes that make them true. What Wittgenstein requires, according to this objection, is that such complexes are not beliefs at all.

To meet the supposed Wittgensteinian requirement, Russell can draw on sense of relations, forms, and an aspect of Russell's ontological categories that we have not yet encountered. Russell claims that if a is a constituent of a complex γ, then γ is not a constituent of a.[24] This claim is about terms standing or not standing in the relation *constituent of*, and so amounts to this: if the complex

$\mathit{Constituent\ of} < a, \gamma >$

exists, no complex is described by

$\mathit{Constituent\ of} < \gamma, a >$

This is akin to the *PoM* view that things never occur as relating relations in a proposition. What such claims amount to is that universals of certain ontological categories only relate terms belonging to certain ontological subcategories.

[24] I follow Giaretta in taking this Russellian claim to rest on an ontological distinction, though he calls it "a difference in type" (1997, 284).

In particular, the relation *constituent of* only relates in the order from entity to complexes in which that entity occurs, never in the reverse order.

Thus, Russell can appeal to facts about the ontological categories of what are related by all the various judgment relations. For example, he might hold that

$$Judge < t_1, t_2; t_3; t_4 >$$

is a belief complex just in case

- t_1 is a subject,
- t_2 is a dyadic relation,
- t_3 is an ordered pair of objects,
- t_4 is the form of dual complexes.

Given these necessary conditions for being a belief complex, (18) doesn't describe a judgment complex because t_2 is Desdemona, who is not a dyadic relation. Russell can grant that (18) is a complex whole, just not a belief complex.

This is a prima facie case that, given forms, sense of relations, and certain general ontological principles, MRT_3 can answer NonP.[25]

2.3.4 Nonconformism

None of this is to say that Russell's logical forms are unproblematic. So far, we have taken forms to be sequences of variables – that is, to be composed of variables in some order. But Russell presents a regress argument against forms being complexes:

- Suppose that forms are complexes; then they have constituents.
- These constituents must be unified in some way to yield the form.
- But forms themselves are supposed to be templates for unifying constituents into complexes.

[25] It may be urged that this prima facie case is open to an objection based on an interpretation of Wittgenstein's January 1913 letter to Russell. In his 1906 substitutional theory, Russell had an explanation of why substitutional simulations of propositional functions are not arguments to themselves, based on what is sometimes called the "grammar" or "syntax" of substitution (for details, see Halimi 2013; Klement 2010; Landini 1998). Perhaps Wittgenstein's letter expresses a demand that "ontological" distinctions be given a similar explanation rather than be taken as primitive. I can't here discuss the complex issues involved beyond noting that the force of this objection depends on a satisfying account of why Russell should find this demand compelling.

- Thus, in order to yield forms from their constituents one needs a form, which in turn has constituents and requires a further form, and so on indefinitely.

If forms don't have structure, then how can acquaintance with a form determine the structure of the truth-making complex of a judgment state?

Russell has an answer to this worry:

> "[S]omething has some relation to something" contains no constituent at all .... In a sense, it is simple, since it cannot be analyzed. At first sight, it seems to have a structure, and therefore to be not simple; but it is more correct to say that it *is* a structure. (*TK*, 114)

A form is what is common to a set of complexes, rather than any constituent part of a complex. An analogy with Frege's doctrine of functions or predicates is useful. A sentential function in *Begriffsschrift* is a pattern common to a number of sentences, and so not a "quotable" part of any sentence.[26]

Even if this answer is cogent, it raises another question. If form is what is common to rather than parts of complexes, then can one be acquainted with a form without being acquainted with complexes having that form? I see no decisive evidence against taking Russell to answer simply no. To be acquainted with a form, we have to be (perceptually) acquainted with a complex of that form. We may need attention to realize that we are acquainted with form in perception, but we must be acquainted with form, else, by the argument of Section 2.3.1, there's no accounting for the difference between understanding and enumerating.

Thus, even if Russellian forms are problematic, it's not clear that they are incoherent. Nor is it clear that Wittgenstein thought so, and thought the MRTs untenable for that reason. For one thing, in *NL*, as we will see in Section 3, Wittgenstein uses a notion of form that seems closely related to, if not the same as Russell's.

So there is no decisive interpretive case for nonsensism.

2.4 Ordinalism

Having cast doubt on the case for form and nonsensism, I turn to the other MRT_3 innovations of *TK*, neutralism, positionalism, and PBs, to explore the prospects of ordinalism.

---

[26] For further discussion of this Fregean view of predicates, see, inter alia, Dummett (1973), Rumfitt (1994), and Noonan (2001, "quotable" appears on 53). The view holds also of sentences expressing generality since for Frege these are formed by filling argument-places of sentential functions with expressions of indefinite indication.

2.4.1 Russellian Neutralism

The chronology indicates that Russell argues for the neutrality of temporal relations and proposes positionalism about them in *TK*, written right after Crit_1. So Crit_1 might be the impetus for these developments.[27]

In *TK*, Russell argues for neutralism with respect to the temporal relation(s) denoted by the terms "before" and "after." He begins by claiming that these terms can be used in distinct descriptions of a single temporal fact:

> ["B]efore" and "after" ... are different in the sense that one cannot be substituted for another in a true statement: if *A* is before *B*, it must not be inferred that *A* is after *B*. But it may be inferred that *B* is after *A*, and it would seem that *this is absolutely the same "fact"* as is expressed by saying that *A* is before *B*. *Looking away from everything psychological*, and considering only the external fact in virtue of which it is true to say that *A* is before *B*, it seems plain that this fact consists of two events *A* and *B* in succession, and that whether we choose to describe it by saying "*A* is before *B*" or by saying "*B* is after *A*" is a mere matter of language. (*TK*, 85; emphases mine)

According to *PoM*, the nonsymmetrical temporal relation denoted by "before" has sense, so that of

(20) x occurred before y

and

(21) y occurred before x

at most one describes an existing complex:

$$Before < x,y >$$
$$Before < y,x >$$

Furthermore, in *PoM*, Russell holds that nonsymmetrical relations have "converses"; indeed, Russell characterizes the "distinction of sense" as "the distinction between an asymmetrical relation and its converse" (*PoM*, Sections 215, 225). The converse of the relation denoted by "before" is the relation denoted by "after"; these mutual converses exhibit the following feature: if x happened before y, then y happened after x, and vice versa.

[27] Ricketts (1996, 69) advances this interpretation. There are at least two other possibilities. First, Russell's neutralism about temporal relations might come from a tacit rejection of McTaggart's (1908) view of the unreality of time. Second, Russell may have worried that temporal relations seem to occur in temporal facts as particularized tropes (see *TK*, 85), the rejection of which from *PoM* onward "represents an essential disagreement with the Hegelians" (Russell 1944, 684). I elaborate on these possibilities in Shieh (forthcoming).

Now, are mutual converses distinct relations? Does (20), if true, describe a different complex from

(22) *y* occurred after *x*?

In *PoM*, Russell holds that the answer is yes: mutual "converses" are distinct. The passages from *TK* just cited display a change of mind. In objective reality, *x* occurred at some time and *y* at another, and these times stand in a temporal relation to one another. We can describe this fact in two equivalent ways, using "before" and "after," but in reality, there are no distinct "converse" temporal relations.

This conclusion leads to trouble for directed temporal relating. Let's use "*T*" for the relation denoted by both "before" and "after." Russell's practice in *PoM* and *PoP* is to take the left-to-right order of occurrence of "*x*" and "*y*" in (20) to indicate that (20) describes the complex

(23) $T < x,y >$

By parity of reasoning, the left-to-right order of occurrence of "*y*" and "*x*" in (22) would indicate that (22) describes

(24) $T < y,x >$

But (23) and (24) are distinct complexes, contradicting Russell's view in *TK*.

Russell considers a way to escape this conclusion, by holding that "it is more proper to go from the earlier to the later term than from the later to the earlier" (*TK*, 87). Then (20) would be what one might call a "privileged expression" of the temporal facts, so that the order of relating is given by the left-to-right order of words in (20); in contrast, the temporal order of relating given by the "non-privileged" expression (22) is the reverse of the left-to-right order of words in (22). Russell rejects this way out, however, on the ground that other "mutually converse" relations, such as "up and down, greater and less, … have obviously no peculiarly 'natural' direction" (*TK*, 87). He concludes, "In a dual complex, there is no essential order as between the terms" (*TK*, 87).

From now on I will occasionally write curly braces around expressions for relata, to reflect Russell's neutralist doctrine. So (20) and (21) describe, respectively,

$T\{x,y\}$
$T\{y,x\}$

The linear order of occurrence of x and y in these expressions does not reflect any difference in how T relates, so what accounts for the fact that (20) and (21) describe different complexes?

Russell has some explaining to do.

2.4.2 Russellian Positionalism

Positionalism is Russell's explanation. Each of (20), (21), and (22) describes a directionless complex. However, speaking metaphorically, in such directionless complexes, there are two "positions" – call them *earlier* and *later* – and each of x and y may "occupy" either of these positions. Both (20) and (22) describe

(25) the complex γ such that x is in the position *earlier* in γ, and y is in the position *later* in γ

while (21) describes

(26) the γ such that y is in *earlier* in γ, and x is in *later* in γ

The difference between the terms "earlier" and "later" lies in a reversal of the "positions occupied" in complexes described using them.

How are we to understand this talk of "occupying positions" in γ? Russell insists that it is *not* as if there are two slots labeled *earlier* and *later* in the relation T, and x and y are fitted into these slots. Such a model for "occupying positions" wouldn't work for *symmetrical* relations such as *next to*. Intuitively, if a is next to b, then b is next to a. Moreover, just like *before* and *after*, a's being next to b is the *same* as b's being next to a. If a's being next to b is like a and b occupying slots α and β in *next to*, then surely there's a *difference* between a's occupying slot α and b's occupying slot β as against b's occupying slot α and a's occupying slot β.

Russell wants to account for the sameness of a's being next to b and b's being next to a by claiming that symmetrical relations such as *next to* have *only one position* for their relata. That would make no obvious sense if this position is like a slot in the relation: what would it mean for two terms to be fitted into the same slot?

But it would make sense if there's nothing more to "occupying a position" than standing in a relation to the complex, for it is perfectly intelligible for two terms to stand in one and the same relation to another term.

So to "occupy a position" in a complex is to stand in a *position relation* to the complex. For example, "x is in *earlier* in γ" really means x stands in the *earlier* relation to γ.

2.4.3 Directionless Believing

Neutralism about temporal relations leads to neutralism about believing, at least for belief complexes whose object-relations are temporal. Consider

(27) Aristotle believes that Socrates is before Plato.

Is this the same belief as

(28) Aristotle believes that Plato is after Socrates?

"Before" and "after" are supposed to denote a single *neutral* relation, *sequence*. According to MRT, it's *sequence*'s relating Plato and Socrates that constitutes the existence of the truth-making complexes. So presumably, "looking away from everything psychological," (27) and (28) describe a single belief about the directionless relation of *sequence*.

But if we accept that

- *believe* relates with sense, and
- the left-to-right order of names of objects in belief-ascription sentences reveals order in believing,

then (27) and (28) denote, respectively,

$$\textit{Judge} < \text{Aristotle}, T; < \text{Socrates}, \text{Plato} >>$$
$$\textit{Judge} < \text{Aristotle}, T; < \text{Plato}, \text{Socrates} >>$$

which are *distinct* judgment states.

So the same Russellian argument leads to the conclusion that if the object-relation of a belief complex is *sequence*, then *judge* relates in no order to form that complex. Russell loses the MRT$_2$ account of the difference between the beliefs described by (27) and by

(29) Aristotle believes that Plato was before Socrates.

This difficulty doesn't stop at "before" and "after." For any expression that, intuitively, denotes a nonsymmetrical relation, there either already is a "converse" or "permutative" expression, or there seems to be no bar to introducing a "permutation": for "above" there's already "below," for "love" there's "is loved by," for "*a* is between *b* and *c*" we could stipulate that it holds just in case "*b* is betwext *c* and *a*," and so on. If such permutative sets of expressions ground arguments for the neutrality of the relations they denote, then the MRT$_2$ solution to DirP goes up in smoke for all beliefs about these relations.

Russell sees not only this, but also that the addition of forms doesn't help:

> In order to understand ["*A* precedes *B*"],we must have acquaintance with *A* and *B* and with the relation "preceding." It [also] requires acquaintance with the general form of a dual complex. But this is by no means enough …; in fact, it does not enable us to distinguish "*A* precedes *B*" from "*B* precedes *A*." (*TK*, 110–11)

If *understand* is neutral, a state of understanding does not order its non-object-relation objects, so naturally the MRT_2 solution to $Prob_{Bel}$ is gone. If all relating is neutral, it's unclear that the variables of forms are ordered, so it's unclear that corresponding complexes with this form can be individuated by an ordering of objects of understanding; the MRT_2 solution to $Prob_{TC}$ is also gone. Even if form variables were somehow ordered, there is no order preserving correlation between objects of understanding and form variables, so the MRT_2 solution to $Prob_{Coord}$ also vanishes.

Evidently Russell formulated TPB in order to provide a solution to DirP without order in believing. Since forms do not aid in the solution of DirP, I will, in discussing TPB, leave them out of account.

2.4.4 Theory of Permutative Beliefs

Suppose *S* has a permutative belief such as that expressed by

(30) *a* loves *b*.

The objects of *S*'s belief seem to be *love*, *b*, and *a*. This leads to $Prob_{TC}$ because from these objects multiple permutative complexes can be formed. It seems (30) describes a complex constituted by $a \rightarrow_{love} b$. If there were order in relating, then this would be different from the complex described by the sentence

(31) *b* loves *a*,

which is constituted by *love*'s relating in the "opposite" direction, $b \rightarrow_{love} a$.

But, according to neutralism, order in relating is an illusion. According to positionalism, the reality underlying the illusion promulgated by (30) is

(32) There is a complex γ such that *a* stands in the position relation C_{lover} to γ and *b* stands in the position relation $C_{beloved}$ to γ,[28]

[28] I have replaced Russell's numerical subscripts on position relations to emphasize that the position relations do *not* replace order of relating by integer order of position relations. Russell's numerical subscripts are merely an indication of identity and difference, not an ordering.

while the reality underlying (31) is (in Russellian notation)

(33) $(\exists\gamma)(bC_{\text{lover}}\gamma \ \& \ aC_{\text{beloved}}\gamma)$

The theory of permutative beliefs is based on the idea that these positionalist explanations – (32) and (33) – *themselves describe complexes*, each of which is uniquely constituted from its constituents. Since these complexes involve the position relations in which constituents of complexes stand to those complexes, I'll call them *position complexes*.

Russell takes the position complexes described by sentences (32) and (33) as the "associated complexes" of the permutative complexes described by sentences (30) and (31). Associated complexes

- are "*unambiguously determined by their constituents*" (*TK*, 145),
- "exist when ever the original [permutative] complexes exist" (*TK*, 145), and
- are "molecular" (*TK*, 147).

Why does Russell think that the associated complexes are uniquely constituted from their constituents? The reason is that position relations differ from nonsymmetrical relations such as *love* in a crucial respect. Constituents of complexes are ontologically distinct from these complexes, so that constituents have positions in complexes, but not vice versa. So, if x is a constituent of γ, and C is a position relation associated with the relating relation of γ, then x stands in C to γ, but γ does not stand in C to x. This means that there is a complex δ constituted by x's being related by C "to" γ, but no complex δ' constituted by γ's being related by C "to" x. In this sense, C is nonsymmetrical. However, C is "heterogeneous" – *not* "homogeneous" – with respect to its constituents in δ, because "interchanging" the terms of δ yields no complex at all, rather than a different complex (*TK*, 123).

Russell thus claims that an associated complex such as (32)

> is non-permutative as regards its *atomic constituents* [$aC_{\text{lover}}\gamma, bC_{\text{beloved}}\gamma$]; also each of these atomic constituents is non-permutative because it is heterogeneous. (*TK*, 147; emphases mine)

The theory of permutative beliefs involves two further moves. After presenting the foregoing account of associated complexes, Russell says, "the permutative complex is not itself the one directly 'corresponding' to the belief," but is, rather, merely "the condition for the existence of the complex which corresponds directly to the belief" (*TK*, 148). What could the latter complex be, except the associated complex? Thus, associated molecular non-permutative complexes are the truth-makers of permutative beliefs.[29]

The second move is not optional, given the basic MRT conception of truth: in the case of belief, "whether a belief is true depends only upon its objects," specifically on whether "the objects are related as the belief asserts that they are" (*TK*, 144). If the corresponding complex of a PB expressible by (30) is the "molecular" complex (32), with atomic constituents $aC_{\text{lover}}\gamma, bC_{\text{beloved}}\gamma$, then surely the objects of this belief are not (merely) *a*, *b*, and *love*, but include C_{lover} and C_{beloved}. For one thing, if the objects were only *a*, *b*, and *love*, then no unique truth-making complex can be constituted from them. Russell puts it this way:

> [I]f I have a *belief whose objects appear verbally to be* [*love*, *a*, *b*], there are *really other objects* ..., and what *I am really believing* is: "There is a complex γ in which [$aC_{\text{lover}}\gamma, bC_{\text{beloved}}\gamma$]." (*TK*, 148; emphases mine)

The theory of permutative beliefs thus distinguishes between *apparent* and *real* objects of PBs, and so also between apparent and real PBs.

It is not altogether clear, from this passage, what exactly are the "real" objects of the "real" belief whose apparent objects are *love*, *a*, *b*. Russell provides a sentence

(32) $(\exists\gamma)(aC_{\text{lover}}\gamma \;\&\; bC_{\text{beloved}}\gamma)$

as "what *S* is really believing," the same sentence that describes the associated complex. Going by this sentence, the belief it indicates is, prima facie, not merely molecular but quantificational. Since Russell hadn't, at this point in *TK*, gotten even to molecular propositional thought, it is a matter for conjecture what he might take to be the objects of *S*'s belief state. It seems hard to deny that *a*, *b*, C_{lover}, and C_{beloved} are among these objects. But what about the variable of quantification "γ"? Since (32) is an existential quantification, one might treat "γ," in its bound occurrences in the scope of the existential quantifier "$(\exists\gamma)$," as a temporary name introduced by existential instantiation. I provisionally assume that it somehow refers to an object of *S*'s belief. Now, in the expression (32) of *S*'s real belief, the position relation C_{lover} appears as predicated of *a* and γ, and the position relation C_{beloved} appears as predicated of *b* and γ. So it would seem that γ is *really* believed by *S* to be a complex

[29] For a complete defense of this claim, on the basis of an extended analysis of the text of *TK*, see Shieh (forthcoming). In taking this passage of *TK* to show that Russell accepted molecular complexes composed of atomic complexes I follow Levine (2013), MacBride (2013), and Pincock (2008).

constituted by the relating relation *love*, and since a and b are believed by S to stand in the position relations associated with *love* to γ, surely a and b are believed by S to be constituents of γ related by *love*. So it would seem that if γ is an object of S's real belief state, then it is a complex $love\{a, b\}$. Since Russell takes the associated complex to be molecular, it seems reasonable to hold that the real belief expressed by (32) involves conjunction in some way. One possibility is that signs such as "&" and "and" denote a logical relation that relates facts. Some commentators hold that, perhaps already by *PM*, Russell rejected such logical relations, but still took conjunction signs to denote a mental object of conjunctive thoughts.[30] I will take conjunction signs to denote an entity *And*, which is a real object of the real belief expressed by (32), without commitment to whether it is a mental or nonmental logical object.

Thus, let's take the real objects of S's belief to be

(34) $a, b, \gamma, C_{\text{lover}}, C_{\text{beloved}}, And$

We now state the TPB solution to DirP.

The answer to Prob_{TC} in the present case depends on the nature of logical objects. Suppose that *And* is an extramental relation of logic. Then, from the objects (34) of PBs expressed by (30) and (31) two *molecular* truth-making complexes, described by

(32) $aC_{\text{lover}}\gamma \;\; \& \;\; bC_{\text{beloved}}\gamma$
(33) $bC_{\text{lover}}\gamma \;\; \& \;\; aC_{\text{beloved}}\gamma,$

may be constituted.

What if "&" denoted an object of thought rather than a logical relation? Both (30) and (31) would then be a description of the existence of a pair of atomic complexes, rather than one molecular complex constituted by the relating of the logical object of conjunction. Distinct truth-conditions consist in the existence of a distinct pair of atomic complexes.

What about Prob_{Bel}? Given neutralism about judging, the obvious Russellian move is to give a positionalist account of the distinction between beliefs expressible by (30) and (31). Such an account would deploy position relations for permutative complexes constituted by *believe* relating a subject S and objects (34). Thus, for example, S's judgment state expressible by (30) is the complex δ constituted by *believe* relating S and objects (34), such that each of these constituents of γ stands in a position relation determined by *believe* to γ. Perhaps something like:

[30] Most prominently Landini (1998, 2014, 2021a). Bostock (2012), Klement (2004), and Stevens (2005) also take Russell not to countenance logical relations by *PM*.

$$(35)\quad (\iota\delta)(SC^B_{\clubsuit}\delta \ \& \ aC^B_{\spadesuit}\delta \ \& \ C_{\text{lover}}C^B_{\text{❦}}\delta \ \& \ \gamma C^B_{\blacklozenge}\delta \ \& \ And C^B_{\heartsuit}\delta$$
$$bC^B_{\flat}\delta \ \& \ C_{\text{beloved}}C^B_{\sharp}\delta \ \& \ \gamma C^B_{\natural}\delta)$$

where $C^B_i, i \in \{\flat, \natural, \sharp, \heartsuit, \blacklozenge, \text{❦}, \spadesuit, \clubsuit\}$, are position relations associated with *judge*,[31] and *And* is the object of belief denoted by conjunction – that is, "and" or "&." If *S* judges (31), then *a* and *b* stand in "reversed" positions to the belief complex:

$$(36)\quad (\iota\delta)(SC^B_{\clubsuit}\delta \ \& \ bC^B_{\spadesuit}\delta \ \& \ C_{\text{lover}}C^B_{\text{❦}}\delta \ \& \ \gamma C^B_{\blacklozenge}\delta \ \& \ And C^B_{\heartsuit}\delta$$
$$aC^B_{\flat}\delta \ \& \ C_{\text{beloved}}C^B_{\sharp}\delta \ \& \ \gamma C^B_{\natural}\delta)$$

As mentioned, *And* may be taken to be either a mental or a nonmental logical object.

The solution to $\text{Prob}_{\text{Coord}}$ is then

- belief complex (35) is true just in case the corresponding molecular complex or pair of atomic complexes (32) exist(s),
- belief complex (36) is true just in case the corresponding molecular complex or pair of atomic complexes (33) exist(s).

2.4.5 The Collapse of Theory of Permutative Beliefs

It is not difficult to see something awry with the TBP solution to DirP. The truth-making complexes of the PBs (32) and (33) are either molecular complexes constituted from distinct sets of atomic complexes, respectively

(37) $aC_{\text{lover}}\gamma, bC_{\text{beloved}}\gamma$

(38) $bC_{\text{lover}}\gamma, aC_{\text{beloved}}\gamma,$

or they are just these distinct pairs of atomic complexes. Assuming that conjunction is symmetric, (32) and (33) are the only molecular complexes that can be constituted from their atomic constituent complexes. However, if we look at the constituents of these atomic complexes, it's obvious that both pairs of atomic complexes are constituted from a single set of entities:

$$a, b, C_{\text{lover}}, C_{\text{beloved}}, \gamma, And$$

We have seen these before; they are the objects (34) of the PBs (35) and (36). That is to say, this one set of objects of both beliefs doesn't determine a unique molecular complex or pair of atomic complexes whose existence would make beliefs with those objects true.

[31] These subscripts are of course chosen to avoid any misleading suggestion of ordering of position relations. My proposed positionalist account of belief derives from Landini (2014).

This conclusion doesn't yet indicate a difficulty for the TPB solution to DirP. This is because TPB individuates beliefs on the basis not only of objects of belief but also on the positions of those objects in belief complexes. But this additional wrinkle doesn't help. Why should the fact that $bC^B_{\spadesuit}\delta$ and $aC^B_{\flat}\delta$ in the judgment complex (36) imply that its corresponding molecular complex or pair of atomic complexes is (33) rather than (32)? Surely it's not because

$$bC^B_{\spadesuit}\delta \ \ \& \ \ C_{\text{lover}}C^B_{\heartsuit}\delta \ \ \& \ \ \gamma C^B_{\blacklozenge}\delta$$

and

$$aC^B_{\flat}\delta \ \ \& \ \ C_{\text{beloved}}C^B_{\sharp}\delta \ \ \& \ \ \gamma C^B_{\natural}\delta$$

occur in the specification of belief complex (36). Position relations are not ordered, so (36) is also specified as

$$(\iota\delta)(SC^B_{\clubsuit}\delta \ \ \& \ \ aC^B_{\flat}\delta \ \ \& \ \ C_{\text{lover}}C^B_{\heartsuit}\delta \ \ \& \ \ \gamma C^B_{\blacklozenge}\delta \ \ \& \ \ AndC^B_{\blacklozenge}\delta$$
$$bC^B_{\spadesuit}\delta \ \ \& \ \ C_{\text{beloved}}C^B_{\sharp}\delta \ \ \& \ \ \gamma C^B_{\natural}\delta)$$

Even if TPB yields cogent solutions to Prob_{TC} and Prob_{Bel}, it fails to solve $\text{Prob}_{\text{Coord}}$.

What if Russell insisted that the "real" objects of *S*'s beliefs *are* the atomic complexes (37) and (38)? Suppose the belief *S* would express by

(30) *a* loves *b*

is false. Then there is no complex γ such that $aC_{\text{lover}}\gamma$ and $bC_{\text{beloved}}\gamma$. Doesn't this mean that the atomic complexes (37) don't exist? If (37) is supposed to be the object of *S*'s belief, then doesn't this mean that *S* in fact has *no* belief expressible by (30)? One could, of course, claim that although complexes don't exist, they nevertheless subsist. But then these subsisting complexes are none other than false Moore–Russell propositions.

Russell himself comes very close to this conclusion. In *TK*, he notes that in order for the associated complex to be non-permutative, its atomic constituents have to be taken as its real constituents, and then he writes:

> [W]hat is more, we have to regard the corresponding propositions as constituents of the proposition "there is a complex γ in which $x_1C_1\gamma, x_2C_2\gamma$, etc." This seems to demand a mode of analyzing molecular propositions which requires the admission that they may contain false atomic propositions as constituents, and therefore to demand the admission of false propositions in an objective sense. This is a *real difficulty*. (*TK*, 154; emphases mine)

"Proposition" in this passage means, not Moore–Russell proposition, but what Russell defines as proposition in *TK*. The definition is based on the idea that a "proposition, in the first place, [i]s something which a number of mental events have in common" (*TK*, 114; for discussion, see Levine 2013). Russell captures this "something in common" by taking a particular subject's state of understanding

$$U(S, x, R, y),$$

and existentially generalizing on the subject S and cognitive relation of understanding U:

$$(\exists U)(\exists S)U(S, x, R, y)$$

(See *TK*, 115; I have omitted Russell's "γ" standing for the form of dual complexes.) The "corresponding proposition" from the last set-out quotation is the existential generalization of some belief state whose objects are the constituents of the associated complex. Russell's conclusion, then, is that if these constituents are atomic complexes, we would have to countenance belief states whose objects are "objectively false propositions." And this is a "real difficulty" since MRTs are supposed to provide relief from false objectives.

The theory of permutative beliefs faces a dilemma.[32] On one hand, if the objects of PBs are the constituents of the atomic complexes that Russell takes to constitute their associated complexes, then more than one set of atomic complexes can be formed from these objects. Even if PBs with the same set of such objects can be differentiated positionally, there is no account of their truth-conditions in terms of these sets of atomic complexes, and so no solution to DirP. On the other hand, if the objects of PBs are the atomic complexes rather than their constituents, then the PBs have different objects and DirP doesn't arise, but objects of false PBs are false objective Moore–Russell propositions, contradicting the rationale for MRTs.

2.5 Is Directed Judging Enough?

The failure of TPB suggests an ordinalist reading of $Crit_2$ as that which ultimately led Russell to realize the failure of TPB, and thereby also the failure of MRT_3 to solve DirP. Wittgensteinian ordinalists see the failure of TPB as serious (see Pincock 2008; Ricketts 1996). Some Russellians take Wittgenstein

[32] Levine (2013), MacBride (2013), and Pincock (2008) also discern a dilemma in case "molecular complex" means a complex constructed by the relating of conjunction. I differ from them in arguing for an analogous dilemma if "molecular complex" means a pair of atomic complexes both of which have to exist for the PB to be true. For more details, see Shieh (2022).

to have played no role in Russellian neutralism and see the failure of TPB as a minor problem for MRTs, easily overcome by simply abandoning neutralism about relating (see MacBride 2013; Lebens 2017). These latter ordinalists share an assumption: the sense of *believe*, perhaps together with forms, succeeds in solving DirP. I now challenge this assumption.

We begin by adopting the assumption that all nonsymmetrical relations relate in various (possibly partial) orders. MRT_2 solves $Prob_{Bel}$ by individuating beliefs in terms of different orders in which *judge* relates objects of belief. It solves $Prob_{TC}$ by individuating corresponding complexes in terms of different orders in which the object-relation relates the remaining objects of belief. It solves $Prob_{Coord}$ by matching the order in which the object-relation relates with the order in which believe relates.

Recall from Section 1.5 that, in *PoP* (198), Russell provides two ways in which order in believing is "indicated":

1. by "order of words in the sentence" ascribing a belief
2. by "inflections, e.g. by the difference between nominative and accusative"

I next show that way 1 doesn't work, and way 2 requires something very much like Moore–Russell propositions.

Before beginning, let's be clear that the issue facing Russell is ultimately not linguistic; it's about whether relations relate terms in various orders. Russell is in general not very interested in language. However, he also claims that "a grammatical distinction" is "*primâ facie* evidence of" "a genuine philosophical difference … and may often be most usefully employed as a source of discovery" (*PoM*, 42). One might say that Russell adopts an attitude of "semantic innocence" (Davidson 1969, 108; for further discussion, see Hochberg 1976, 23–4; Hylton 1990, 171; Turnau 1991, 53–4; Levine 2004, 262–9, 281). Russell's two linguistic ways of "indicating" order in relating is an instance of this attitude. The problems I will articulate for these ways then poses a challenge to Russell: if there is order in relating, what exactly is that order, and how exactly is it expressed in language?

2.5.1 "Deviant" Belief Ascriptions and Word Order

I begin the argument against way 1 with a line of thinking leading to neutralism about belief slightly different from that in Section 2.4.3, and closer to Russell's own route to neutralism about temporal relating discussed in Section 2.4.1.

Especially in English poetry,

(6) Cassio Desdemona loves

may be used to say what is said by

(11) Desdemona loves Cassio

(See Ferber 2019, 105–9.) But then, would not

(5) Othello believes that Cassio Desdemona loves

ascribe the *same belief* to Othello as

(1) Othello believes that Desdemona loves Cassio?

Were Emilia to utter (1), and then at a later time (5), to Bianca, would Bianca not reasonably conclude (if she trusts Emilia) that Othello has not changed his mind about whether Desdemona loves Cassio?

However, the order of the words "Desdemona" and "Cassio" in (5), whether left-to-right or right-to-left, is the "reverse" of the order of "Desdemona" and "Cassio" in (1). So, however the order of believing is "indicated" by the order of words, it's hard to see how believing would relate Cassio and Desdemona in the *same* $\rightarrow_{believe}$ *order* in the complexes described by (1) and (5). It follows, then, that (1) and (5) describe *distinct belief states.*

Now, one response to Russell's argument for temporal neutralism is Russell's own in *PoM* (Section 219): insist that each nonsymmetrical relation is distinct from its converse, so that

A is before *B*
B is after *A*

describe distinct temporal complexes, but it's a "brute fact" that whenever either exists so does the other (see, e.g., MacBride 2007, 55).

So, to maintain directionalism for *believe*, one might bite the bullet and insist that, appearances to the contrary, (1) and (5) ascribe different beliefs to Othello and describe distinct belief complexes. It's just a "brute fact" that whenever any subject has one of these beliefs, they also have the other.

However, *believe* differs from temporal sequence relations in a key respect: beliefs have truth-values. So, even if one is willing so to bite the bullet, it seems hard to deny that these distinct beliefs are either both true or both false. Indeed, this constancy of truth-value is of a piece with Bianca's intuitive reaction that, going by Emilia's utterances, Othello has not changed his mind.

According to MRT_2, in the truth-making complex, the object-relation relates in the same order as *believe* relates the other objects of belief in belief complexes. Hence, given that believing relates Cassio and Desdemona in different, "reverse" $\rightarrow_{believe}$ orders in beliefs (1) and (5), in their corresponding complexes *love* relates in "opposite" directions. Indeed, accepting our doubly de-fictionalized Shakespeare as fact, one of (1) and (5) is true and the other false.

In the face of this problem for the bullet-biting response, one might unclench the bullet and consider formulating rules governing the indication of order of believing from the order of words in belief ascriptions that map both (1) and (5) to the same belief state, or at least belief states in which Cassio and Desdemona are in the same $\rightarrow_{believe}$ order.[33]

Perhaps, for example, if in a complement clause the verb denoting the object-relation occurs last in the left-to-right word order – as in that of (5):

(6) Cassio Desdemona loves

then the order of the object-relation's relating in the truth-making complex is the reverse of the left-to-right orders of the names of the other objects of belief:

(7) *Love* < Desdemona, Cassio >

Of course, more complex rules would have to be formulated for three and more place nonsymmetrical relations –

Roderigo to Cassio Desdemona commends

But we can see that the current proposal already runs into trouble by considering a Latin translation of ascription (1):

(39) Othello credit Desdemonam Cassium amāre

In the complement clause of this ascription,

(40) Desdemona Cassium amat[34]

the verb occurs left-to-right last, so according to the contemplated rule, the corresponding complex is

(8) *Love* < Cassio, Desdemona >,

which, in contrast to (7), exists.

---

[33] I'm grateful to Vera Flocke, Kirk Ludwig, and Katy Meadows for pressing this reply, and to Tyke Nunez for the following specific account.

[34] Readers with Latin will have noticed that I have tacitly set aside the complication that in complement clauses of belief ascriptions such as (39) both names would usually appear in accusative case, so that this displayed sentence is not strictly the complement clause but the declarative sentence corresponding to it.

2.5.2 Inflections

Way 2 holds that in an inflected language such as Latin, the fact that "Desdemona" occurs in the nominative and "Cassium" in the accusative cases in the complement clause (40) of (39) indicates that Desdemona$\rightarrow_{judge}$Cassio. The grammatical difference between nominative and accusative corresponds to the difference between grammatical subject and direct object in English. Thus, according to way 2, that in

(11) Desdemona loves Cassio

"Desdemona" occurs grammatically as subject and "Cassio" as direct object also "indicates" the same order of believing. "Deviant" ascriptions no longer pose a difficulty because in, for example,

(6) Cassio Desdemona loves

"Desdemona" occurs as subject and "Cassio" as direct object.

According to way 2, then, because

- In both (11) and (6) "Desdemona" occurs as subject and "Cassio" as direct object, and
- "Desdemona" occurs in nominative and "Cassium" in accusative in (40)

all of ascriptions (1), (5), and (39) describe Desdemona$\rightarrow_{judge}$Cassio.

But why this order? Why not direct object or accusative $\rightarrow_{judge}$ subject or nominative? One answer is that "Desdemona's" occurring grammatically as subject (in nominative) in a sentence whose main verb is "love" ("amare") implies that this name does not *merely* refer to Desdemona, but it also does something else. It also somehow "indicates," to use Russell's word in *PoP*, that this referent *is a lover*. Similarly, that "Cassio" occurs as direct object of "love" (in an accusative with "amare" as main verb) implies that this name refers to Cassio, and in addition "indicates" that he *is a beloved*.[35]

Now, what exactly is it for an expression to "indicate" that an object *x* "is a lover"? Whatever exactly this "indication" is, isn't indicating that *x* is a lover indicating that *x* has the property of being a lover? What else could that be except to indicate that *x* stands in the relation *love* to some object?

[35] I'm grateful to Paul Howatt and Gary Ebbs for suggestions leading to this proposal.

That is to say, to be "indicated" as being a lover is to be "indicated" as being a constituent of a complex in which *love* relates "from" it "to" something else. What if, in fact, *x* doesn't stand in the object-relation to anything? Then this account of being indicated as lover is committed to the existence of an objectively false complex containing *x*.

If for *x* to be "indicated as being a lover" doesn't mean for *x* to be indicated as having the property of being a lover, what does it mean? Perhaps it means that *it is as if x* has this property. But what does that mean? Is it for *x* to be believed to have this property? What could this mean except that the object of belief isn't merely *x* by itself, but *x* standing the relation *love* to something? Again, this account of "indicated as lover" doesn't seem to escape the potential for false complexes that Russell sought the MRTs to avoid.

2.6 Interpretive Speculation

Let's take stock.

In Section 2.3, I showed that if nonsymmetrical relations have sense, and if, in particular, we can take for granted that believing relates in various orders, then, by assuming acquaintance with forms, Russell has solutions to both DirP and NonP. This suggests that, if Wittgenstein's criticisms are compelling, they are something other than what nonsensists take them to be.

In Section 2.4, I argued that Russell's positionalist attempt to solve DirP, once he came to reject directionalism about relating, fails. The theory of permutative beliefs succeeds in solving Prob_{Bel} by distinct positions in belief complexes, and in solving Prob_{TC} by distinct atomic position complexes. But it fails to solve $\text{Prob}_{\text{Coord}}$. Hence, there is no account of how PBs with the same "real" objects are determined as true or false by the existence of complexes constituted from these objects. There would be an account if the "real" objects of PBs are atomic position complexes, but this opens the door to the existence of false complexes.

In Section 2.5, I argued that Russell would be no better off if he accepted directionalism. The difficulty here is Prob_{Bel}. It's supposed to be solved by distinct orders of believing. But the phenomenon of "deviant" belief ascriptions shows that, given a set of objects of belief, it is not clear how they are ordered by the relation of believing. If there's no answer to Prob_{Bel}, there is, a fortiori, no account of how beliefs with the same objects have distinct truth-values. One could ground orderings of objects by believing by following up Russell's sense that these orderings are indicated by grammatical features of belief ascriptions. But the account seems also to invite the existence of false complexes.

Russell is thus left with no solution to DirP. If, as I argued in Section 2.3, a path to blocking substitution argument that engenders NonP requires directionalist believing, he's also left with no answer to NonP.

Note now two shared features of the difficulties for TPB and for sense in believing:

- One problem in each case is that the MRT fails to explain how beliefs have determinate truth-values.
- In each case, positing complexes constituted from objects of belief overcomes this problem.

These features cohere with Wittgenstein's June 1913 "exact" statement of what is wrong with Russell's theory of judgment:

> from the proposition "A judges that (say) *a* is in a relation *R* to *b*," … the proposition "$aRb.\vee.{\sim}aRb$" must follow directly without the use of any other premises.

As many commentators have noted, the significance of "$aRb.\vee.{\sim}aRb$" following from "A judges that *aRb*" may be that A's belief is either true or false – that is, has a truth-value. Wittgenstein would then be characterizing his objection as the claim that Russell's MRTs fail to explain how beliefs have determinate truth-values. The "other premises" Wittgenstein alludes to may then be understood as descriptions of complexes by the addition of which Russell's accounts do achieve determinate truth-values for beliefs. For Wittgenstein's case, these would be the "premises" "$aC^R_*\gamma$" and "$bC^R_\dagger\gamma$" to TPB, and the "premises" "$(\exists x)aRx$" and "$(\exists x)xRb$" to MRT_2 to fix $a \rightarrow_J b$. Note that these premises are "other" in the sense that they are contrary to MRTs, and not merely not part of Russell's MRT.

This takes me to an interpretive hypothesis. What Wittgenstein's $Crit_2$ showed Russell is:

> Unless some objects of belief are complexes constituted by object-relations relating other objects of belief, the MRTs do not furnish beliefs with determinate truth-values.

As stated, it's not clear how such objects of belief are different from Moore–Russell propositions. Positing such objects is not yet to be committed to a DRT, since, as we will soon see, one need not hold that judgment is a relation to one such entity. However, it's clear that if false beliefs involve such objects, then some of them are liable to be false complexes. Prima facie, there are at least two options:

1. Show that none of these objects are false complexes.
2. Find complexes constituted by relating of relations other than the object-relation that somehow play a role in determining truth-values for beliefs.

Here's evidence that Russell took option 1 until 1919 and Wittgenstein in *NL* took option 2. In Section 3, we'll examine Wittgenstein's option 2 in detail.

1. In "Props," written possibly right after Crit_2, Russell contemplates postulating "neutral facts" in an MRT to "replace[] form" (1913a, 197). A neutral fact composed of "objects *x*, *R*, *y*" is "*a relation of x and R and y*," which exists "whenever these objects" "form one or other of two complexes *xRy* or ~*xRy*" (195; emphases mine). The important point is that a neutral fact, whatever exactly it is, involves the object-relation *R* relating *x* and *y*. But its neutrality means that it's neither the "positive fact" *xRy* nor the "negative fact" ~*xRy* but "a *constituent* of the positive or negative fact" (195); hence no neutral fact is a false complex.
2. After apologizing for "paralysing" Russell, Wittgenstein in July 1913 tells Russell, "I think [my objection to your theory of judgment] can only be removed by a correct theory of propositions" (*WC*, 42).
3. Wittgenstein attempts such a "correct theory of propositions" in *NL*. As we will see in Section 3, in *NL*, propositions are facts, or rather, essentially involve facts. But a propositional fact is not composed of the same entities as those entities whose failure to form a fact falsifies the propositional fact.
4. In response to Russell's queries about *NL*, Wittgenstein writes in November 1913, "the symbol for 'a fact' is a prop[osition] and this is no incomplete symbol" (*WC*, 50).
5. In 1918, in *PLA*, Lecture IV, Russell claims that a belief complex has two relating relations, in particular the "subordinate verb" occurs in it "as relating," even in a "false belief":

> [T]he verb "loves" occurs in th[e] proposition ["Othello believes that Desdemona loves Cassio"] and seems to occur as relating Desdemona to Cassio whereas in fact it does not do so, but yet it does occur as a verb, it does occur in the sort of way that a verb should do. [T]he subordinate verb (i.e., the verb other than believing) is *functioning as a verb*, and *seems to be relating two terms*, but as a matter of fact *does not when a judgment happens to be false*. That is what constitutes the puzzle about the nature of belief. (*PLA*, 59; emphases mine)

Russell still adheres to an MRT, but understands the moral of Wittgenstein's criticism to be that the object-relation has to "function as a verb." It's not clear what this could mean except that it has to be relating the other objects of belief. But of course if the belief is false, the object-relation doesn't so relate. Russell dodges a plain contradiction by holding that the subordinate verb "seems to be relating." But it's not clear that he has an explanation of what this means. The puzzle posed by this two-verb MRT, as I'll call it, is not really a matter of there being two relations in a cognitive state, but rather that one of the two seems both to have to relate and also not to relate.

6. Finally, in "On Propositions," Russell lets go of MRTs and holds that propositions are complex entities constituted from mental images or words (Russell 1919b, hereafter *OP*, 29).

The exegetical speculation I just proposed is a modified Wittgensteinian ordinalism. I accept, following Russellian ordinalists, that Russell may have come to neutralism on his own steam. However, I side with other Wittgensteinian ordinalists in taking Wittgenstein's $Crit_2$ to be a significant threat to the MRTs. But the threat doesn't depend on neutralism. Even accepting that relations have sense, Russell had no clear solution to DirP consistent with MRTs. Appeal to some sort of complexes seems required for determinate truth-making for beliefs.[36]

Before going on, I address a salient issue for my reading. What of the apparently overwhelming evidence in favor of nonsensism? What, in particular, of the fact that in the *Tractatus*, Wittgenstein charges "Russell's theory" with failing to show that "it is impossible to judge a nonsense"? Note to begin with that, according to the *Tractatus*, if a propositional sign "has no sense, this can only be because we have not given some of its constituents a meaning" (5.4733). Nonsense is not the result of some sort of "clash" of meanings as suggested by Russell's ontological doctrine that things do not play the role of relating relations.[37] Whatever exactly a Tractarian piece of nonsense is, it is not determinately either true or false. That's exactly the problem with the MRTs. From this perspective, we see an underlying unity to NonP and DirP: they are different ways in which the MRTs fail to provide determinate truth values for beliefs. The nonsense problem arises where there are *no* complexes constituted by the objects of a purported judgment complex, and so a fortiori no truth-making complexes. The direction problem, arises where multiple

[36] To answer the title of Griffin (1991): Russell fell, but was then shot before he could get back up.

[37] This is especially emphasized by Conant (2002) and Diamond (1991).

logically possible complexes can be composed from the objects of a belief, so there is no determinate answer as to which makes the judgment true.

3 "Notes on Logic"

The "correct theory of propositions" undertaken in *NL* evidently should at least steer a course between the Scylla of MRT's DirP and the Charybdis of DRT's problem of falsity.

In the following, I refer to *NL* in the form *x-y* for manuscript *x*, remark *y*, and *x* = S for Russell's Summary.

3.1 Facts, Positive and Negative; Sense; Meaning; Negation

Wittgenstein reinstates propositions as elements of belief. But propositions are made up of different things from what they represent – that is, there is a medium of representation. In this section, I outline basic features of *NL* propositions and some salient puzzles they present.

Propositions "are *symbols* having reference to facts," and, more important, "are *themselves facts*: that this inkpot is on this table may express that I sit in this chair" (*NL*, 1–2; emphases mine).

Propositions have *sense*:

> Every proposition is essentially true-false: to understand it, we must know both what must be the case if it is true, and what must be the case if it is false. Thus a proposition has two *poles*, corresponding to the case of its truth and the case of its falsehood. We call this the *sense* of a proposition. (*NL*, S–13).
>
> I understand the proposition "*aRb*" when I know that either the fact that *aRb* or the fact that not *aRb* corresponds to it. (*NL*, 4–6)

Apparently the sense of a proposition "*aRb*" is a *pair* of facts – the fact that *aRb* and the fact that not *aRb* – and these facts are the *poles* of the proposition.

Moreover, propositions have *meaning*:

> [W]e can only know the *meaning* of a proposition when we know if it is true or false. (*NL*, 3–25; emphases mine)[38]

Meaning is somehow connected with *negation*:

> In my theory *p* has the *same meaning* as not-*p* but *opposite sense*. The meaning is the fact. (*NL*, S–20; emphases mine; see also 4–2)

[38] "Meaning" is plausibly Wittgenstein's or Russell's translation of the German *Bedeutung*, since in "Notes Dictated to G. E. Moore" in April 1914, we read "The *Bedeutung* of a proposition is the fact that corresponds to it" (Wittgenstein 1979, hereafter *NB*, 112).

> [I]t is important that we *can* mean the same by "*q*" as by "not-*q*," for it shows that neither to the symbol "not" nor to the manner of its combination with "*q*" does a characteristic of the denotation of "*q*" correspond. (*NL*, 1–9)

Wittgenstein also talks of *positive* and *negative facts*:

> There are positive and negative facts: if the proposition "this rose is not red" is true, then what it signifies is negative. But the occurrence of the word "not" does not indicate this unless we know that the signification of the proposition "this rose is red" (when it is true) is positive. It is only from both, the negation and the negated proposition, that we can conclude to a characteristic of the significance of the whole proposition. (*NL*, 1–7).

As we saw in Section 2.6, in "Props," possibly written right after Crit_2, Russell uses the idea of negative facts; as far as I know, this is the first appearance of this notion in any text of Wittgenstein or Russell.

It's less than clear what this all means. Here are some obvious questions:

- The fact that *aRb* is a positive fact; is the fact that not-*aRb* negative?
- What is a negative fact?
- How is it that a proposition and its negation have the same meaning but opposite sense?
- It seems Wittgenstein is saying this: whether what corresponds to a given proposition, if true, is a positive or a negative fact is not determined independently of whether the negation of that proposition, if true, corresponds to a positive or a negative fact. Why?

3.2 The Symbolizing of Forms of Propositions

To resolve these puzzles, I turn to a well-known and perhaps even more puzzling text from *NL*:

> [T]he form of a proposition symbolizes in the following way: Let us consider symbols of the form "*x*R*y*"; to these correspond primarily pairs of objects, of which one has the *name* "*x*," the other the *name* "*y*." The *x*'s and *y*'s stand in various relations to each other, among others the *relation* R holds between some, but not between others. I now determine the sense of "*x*R*y*" by laying down: when the facts behave in regard to {editors' note: *sich verhalten zu*, are related to} "*x*R*y*" so that the *meaning* of "*x*" stands in the relation R to the *meaning* of "*y*," then I say that the[y] [the facts] are "of like sense" ["*gleichsinnig*"] with the proposition "*x*R*y*"; otherwise, "of opposite sense" ["*entgegengesetzt*"]; I correlate the facts to the symbol "*x*R*y*" *by* thus dividing them into those of like sense and those of opposite sense. To this correlation corresponds the correlation of name and meaning. Both

> are psychological. Thus I *understand* the form "*x*R*y*" when I *know* that it discriminates the behaviour of *x* and *y* according as these stand in the relation R or not. (*NL*, 4–8; emphases mine; all square brackets in original)[39]

This passage points to some central claims of Wittgenstein's theory:

- Facts have *forms*.
- A fact represents through the *symbolizing of (one of) its form(s)*.
- The symbolizing of a form involves a *stipulation* (a "laying down").
- The stipulation "determines" the *sense* of a proposition of a form by "dividing" or "discriminating" facts as "like" or "opposite" sense with the proposition.
- Understanding a proposition with a form is or involves "knowing" how it divides or discriminates facts.

I now turn to an interpretation of these claims.

The facts that are "symbols" or "propositions" are *aspects* of a collection or a composite. The form of a propositional fact is something that it has in common with other propositional facts. Let's turn to some examples.

First, a drawing:

Figure 1

There are many facts about Figure 1, for example:

1. the (image of the) panda eating the apple is to the (audience-)left of the panda not eating
2. both pandas have a black ear
3. there is grass behind the pandas, and grass in front of them.

Keep in mind fact 1.

[39] Readers observant of distinctions between use and mention would likely be at least puzzled by this text. In Shieh (forthcoming), I show how the puzzlement can be overcome.

Second, a sequence of Latin alphabet letters and blank spaces:

(41) Desdemona loves Othello

Here's one among many facts about it:

4. The word "Desdemona" occurs to the left of the word "loves," and the word "Othello" occurs to the right of "loves."

Third, a sequence of characters:

苔
丝
狄
蒙
娜
爱
奥
赛
罗

One among many facts about this sequence is:

5. The sequence of characters 苔
丝
狄
蒙
娜 occur above 爱 and the sequence of characters 奥
赛
罗 occur below 爱

Here's a way of characterizing facts 1, 4, and 5:

- One panda(-image) stands in the *left of* relation to the other panda.
- The word "Desdemona" stands in a *complex horizontal spatial relation*, a relation involving the word "love," to the word "Othello."
- A vertical sequence of five characters stands in a *complex vertical spatial relation*, a relation involving the character "爱,"to a vertical sequence of three characters.

So conceived, something they have in common shows itself:

Some entity stands in some relation to another entity

This is the *form* of these facts; it is what Wittgenstein indicates in *NL*, 4–8 with

*x*R*y*

The allusion to Russell's specification of the form of dual complexes (*TK*, 98) leaps to the eye; hence, what our sample facts have in common can also be conveyed by Russell's characterization of this form: "Something has some relation to something" (*TK*, 114).

Wittgenstein holds that the form of a proposition "symbolizes." I elaborate in terms of the drawing of the pandas. Recall again fact 1 about this photograph:

(42) The panda eating the apple is to the left of the panda not eating

The form of fact (42) is *x*R*y*, so, we can use it to express any fact that has this form.

This use rests on "determin[ing] the sense of '*x*R*y*' by laying down" what facts are of like sense and what facts are of opposite sense. To pick out a specific fact for fact (42) to express, there has to be *rules* that fix

- With what entity the left-side panda is correlated (this entity is the "*meaning*" of the left-side panda treated as a *name*),
- With what entity the right-side panda is correlated (the *meaning* of the right-side panda as name),
- What fact about these meanings makes fact (42) *true* – this Wittgenstein calls a stipulation of which "behavior" of the facts is "of like sense" with (42) – and
- What fact makes fact (42) false, that is to say, "behavior" of the facts is "of opposite sense" with (42).[40]

[40] The reader may at this point wonder whether the account in the text is consistent with 4–8 – "I correlate the facts to the symbol '*x*R*y*' *by* thus dividing them into those of like sense and those of opposite sense" – which suggests that a stipulation of sense classifies *every fact*, or *every fact about the meanings*, as like or opposite sense with a propositional fact. I cannot go into this issue in detail here, but here's one consideration against the suggestion. If a stipulation has to classify all facts, then it would have to say whether the fact that Othello loves Desdemona is of like or opposite sense to the propositional fact "Desdemona loves Othello." On one hand, it seems hard to accept that if it is the case that Othello loves Desdemona, then "Desdemona loves Othello" is true. So we wouldn't want to stipulate that this fact is of like sense with "Desdemona loves Othello." On the other hand, it seems equally hard to accept that if it is the case that Othello loves Desdemona, then "Desdemona loves Othello" is false. So we wouldn't want to stipulate that this fact is of opposite sense with "Desdemona loves Othello" either.

These *rules* specify *how* a propositional fact is *compared with* facts (*NL*, S–18: "A proposition is a standard to which facts behave"). The rules are *conventionally adopted* or *stipulated* (for "convention," see *NL*, S–15, n2, 3–14). Rules of comparison determine the sense of a propositional fact, so I'll call stipulations of such rules *sense-stipulations*.

Propositions are *not only* facts; they are facts *together with* rules of comparison.[41]

Although, as I formulated them, the rules of comparison for (42) correlate each panda with its meaning, *none* of them explicitly *correlates* the relation *left of* in the panda fact (42) *with a relation as its meaning*. This is meant to reflect Wittgenstein's silence, in *NL*, 4–8, on whether a relation that holds of entities to constitute a propositional fact symbolizes through correlation with a meaning, and if so whether that meaning is a relation.

What underlies this silence is this. The symbolization of propositional facts does require that their relating relations, to use Russell's terminology, be correlated with relations. But correlation of relation with relating relation works differently from correlation of meaning with name. All there is to naming is stipulation of an object. But a relation *R* correlated with a propositional relating relation either holds of a given ordered sequence of entities of the same length as the number of argument places of *R*, or it doesn't. That is to say, either *R* and this order sequence of entities constitute a positive fact, or they constitute a negative fact. Correlating with *R* is correlating all such "behaviors" of *R* and sequences of entities. Together with the meanings of the names in the propositional fact, such a correlation with *R* fixes whether these meanings stand in *R* in some order, or not.

These stipulated correlations are not enough to account for the symbolizing of propositional facts. Correlating *left of* with some two-place *R*, for instance, does not settle whether panda fact (42) is to describe that *R* holds of the meanings of the pandas, or that *R* does not hold of them. It is only by fixing which of these is "of same sense" and which is "of opposite sense" with (42) that it is determinate what (42) is describing, and so what proposition it is.

For example, say we want to use facts involving *left of* to represent facts involving *love*. This requires a rule specifying, for any pair of objects *x*,*y*,

- whether the positive fact that *x loves y* makes a fact that *a* is to the left of *b true*, when *x* is stipulated to be the meaning of *a* and *y* the meaning of *b*, *and*

[41] Russell (1906a, 185–6) experiments with a theory on which the object of a belief consists of ideas related in some way, which seems similar to the *NL* conceptions of proposition, and also the view of Russell (1919), discussed in Section 7.1.

- whether the negative fact that *x does not* love *y* makes a fact that *a* is to the left of *b false*, when *x* is stipulated to be the meaning of *a* and *y* the meaning of *b*.

Let's say that here the positive fact is stipulated as truth-making, the negative fact as falsity-making. I would like to stress that the positivity of a fact does not imply that it is truth-making, nor does negativity imply falsity-making. We will see that this point underlies the *NL* theory of negation.

Here is a sample sense-stipulation, (🐼), for (42):

- the apple-eating panda means Desdemona,
- the panda not eating means Othello, and
- the positive fact that the meaning of the left-side panda *loves* the meaning of the right-side panda makes fact (42) true.
- the negative fact that the meaning of the left-side panda *does not love* the meaning of the right-side panda makes (42) false.

Together with these rules, the fact (42) about *images* in a drawing and the relation *left of* is a *proposition*; call it "Pandas." Pandas expresses or represents the Shakespearean fact that Desdemona loves Othello, about *people* and the relation *love*.

Has the *NL* theory skirted Charybdis? The answer turns on whether the notion of negative fact is coherent, or, at any rate, less problematic than the Moore–Russell conception of a unified fact-like entity having the unanalyzable property of falsity. I suspect that two factors made the idea acceptable to Wittgenstein at the time of *NL*. First, a statement like "in fact the dog didn't bark," seems perfectly coherent. This suggests that there are *facts* about what is *not* the case. Second, the dog's failure to bark seems precisely an *absence* of a state of affairs in which the dog is "unified" with barking. Such facts of absences aren't entities "in the world … which can be described as objective falsehoods," the existence of which Russell found "almost incredible" (*NTF*, 176).[42] But, as we will see, a mere year or so after *NL*, Wittgenstein came to doubt the coherence of negative facts.

[42] Russell entertained and at times defended negative facts during the period from "Props" (1913a) through *The Analysis of Mind* (1921); one such defense, he reports in *PLA* (42), "nearly produced a riot" in his 1914 Harvard audience. At one point he argues that negative facts are not absences of facts: "the absence of a fact is itself a negative fact; it is the fact that there is *not* such a fact as A loving B" (*OP*, 5). This is not inconsistent with the view I've hypothesized for Wittgenstein, for what Russell rejects is the elimination of negative facts rather than their construal as facts about absences. For the "near riot," see Linsky (2018). For further discussion of Russell's not altogether firm commitment to negative facts, see Perovic (2018).

3.3 Negation

I present the *NL* theory of negation using the second fact of form "*x*R*y*" of Section 3.2, about the sequence of Latin letters and blank spaces,

(41) Desdemona loves Othello

The fact is:

(43) The word "Desdemona" occurs to the left of the word "loves," and the word "Othello" occurs to the right of "loves"

Here is sense-stipulation, call it (†), for fact (43):

- "Desdemona" means Desdemona.
- "Othello" means Othello.
- The positive fact – call it F^+ – that the meaning of the name on the left of "loves" *loves* the meaning of the name on the right of "loves" makes fact (43) true.
- The negative fact – called F^- – that the meaning of the name on the left of "loves" *does not love* the meaning of the name on the right of "loves" makes (43) false.

(†) surely reflects an ordinary English meaning of (41). So let's say that (†) is a *conventional sense-stipulation* for fact (43). Let's say that each of the facts F^+ and F^- is the *opposite* of the other. If F^+ obtains, then its opposite F^- doesn't, and vice versa.

Wittgenstein's account of negation rests on the intuitive idea that if a proposition is true, then its negation is false, and vice versa. This suggests that the negation of a proposition *p* should be a proposition *q* satisfying the following condition:

q is of like sense with the fact that is of opposite sense with *p*, and *q* is of opposite sense with the fact that is of like sense with *p* (*)

I now describe a sense-stipulation, (‡), for (43) differing from (†) in:

- the negative fact F^- makes (43) true
- the positive fact F^+ makes (43) false.

(‡) reverses the like-sense and opposite-sense facts of stipulation (†). Let's say that these two *stipulations*, (‡) and (†) are "of opposite sense" to one another.

Now suppose that *p* is the proposition consisting of the fact (43) together with (†). Then the condition (*) for being the negation of *p* is satisfied by the fact (43) together with stipulation (‡). That is to say, fact (43) together with stipulation (‡) *is* the negation of the proposition that consists of this very fact together with the conventional stipulation (†).

In most natural languages, negation is often formed by adding some expression of negation to a sentence – "not," "不,"and so forth. So we normally take the negation of (41), considered as an English sentence, to be

(44) Desdemona does not love Othello

But from Wittgenstein's perspective, what is important is the relationship of opposition between sense-stipulations rather than the propositional facts. So "not" is not required for negation.

Moreover, the occurrence of the word "not" in (44) does not imply that a fact about it is made true by the negative fact F^-, unless a suitably related fact about (41) is made true by the opposite positive fact F^+. Nothing stands in the way of giving (41) the sense-stipulation (‡), opposite in sense to the conventional stipulation (†), and (44) the conventional (†). In this case, we might say that (41) is the negation of (44), formed by omitting "not" from the propositional fact (44).

3.4 Judgment and DirP

As we saw, in *NL*, 4–8, Wittgenstein claims that understanding the form "*x*R*y*" requires "knowing" how it "discriminates" positive and negative facts involving pairs of objects standing or not standing in the relation correlated with "R." This suggests an account of understanding propositions: it is, for example, to know the rules of a sense-stipulation comparing a propositional fact of form "*x*R*y*" with facts in the world.

In making a judgment or forming a belief, a subject is not related to a single Moore–Russell propositional object. Nor is the subject related to a multiplicity of objects, as in the MRTs. The thinker is "related to" a (propositional) fact together with rules of comparison with facts in the world. As Wittgenstein says, "This is obviously not a relation in the ordinary sense" (*NL*, S–22).

In order for the *NL* theory to address DirP, it would have to account for the difference between, for example,

(45) Iago believes that Desdemona loves Othello
(46) Iago believes that Othello loves Desdemona

The extraordinary nature of the *NL* belief relation shows itself in the fact that there are *many ways* for (45) to be true of Iago. One would be for Iago to take the panda fact (42) to be compared with the world in accordance with the sense-stipulation rules (🐼) of Section 3.2. This would be for Iago to have the belief in question by being (extraordinarily) related to Pandas.

This is only one among many ways for Iago to have this belief, because there are many facts of the form *xRy* that can be compared to worldly facts in ways analogous to (🐼) For instance, fact 3 from Section 3.2 may be paired with these rules:

- The character sequence 苔丝狄蒙娜 means Desdemona,
- The character sequence 奥赛罗 means Othello, and
- The positive fact that the meaning of the character sequence above 爱 *loves* the meaning of the character sequence below 爱 makes fact 3 true.
- The negative fact that the meaning of the character sequence above 爱 *does not love* the meaning of the character sequence below 爱 makes fact 3 true.

Call the proposition consisting of fact 3 and this sense-stipulation 爱.

What now about Iago's belief described by (46), according to which, as Russell might put it in *TK*, the positions of Desdemona and Othello with respect to *love* are reversed? Call this the *reverse belief.* The answer evidently depends on the propositional fact and the sense-stipulation that the reverse belief involves.

Suppose that Iago has belief (45) with Pandas. One way he could have belief (46) is to keep fixed the propositional fact (42), but change the sense-stipulation from (🐼) to (🐼′):

- the apple-eating panda means Desdemona,
- the panda not eating means Othello, and
- the positive fact that the meaning of the right-side panda *loves* the meaning of the left-side panda makes fact (42) true.
- the negative fact that the meaning of the right-side panda *does not love* the meaning of the left-side panda makes (42) false.

Here the original rule of comparison (🐼) is altered by *reversing* the meaning of the left-side panda and the meaning of the right-side panda in the truth- and falsity-making facts. Call the proposition that is (42) together with (🐼′) Pandas′.

A different way for Iago to have the reverse belief is to change to a different drawing,

Figure 2

and a propositional fact about Figure 2:

(47) The panda not eating is to the left of the apple-eating panda.

One might say that (47) "reverses the pandas": the panda to the left is now to the right, and vice versa. But then we keep sense-stipulation (🐼) (again see p. 55). Call the proposition that is (47) together with (🐼) Pandas″.

Note that Iago may make the judgment (45) through either Pandas or 爱, which are evidently different propositions. Similarly, Iago might make the "reverse" judgment (46) through either Pandas′ or Pandas″, which are also different propositions. This indicates a salient difference between *NL* and Moore–Russell propositions: *NL* propositions are *not the objects of belief*.[43]

Has Wittgenstein cleared Scylla? Here is a way of seeing how the *NL* theory of belief dissolves DirP. "Notes on Logic" replaces the unified entities that are Moore–Russell propositions with representing facts. "Notes on Logic", as it were, builds into propositions *comparison*, not only to the *truth-making* complexes of the MRTs, but also to *negative* falsity-making complexes. As we read in S–13, to understand a proposition, "we must know both what must be the case if it is true, and what must be the case if it is false." This blocks the problem of fixing truth-conditions that felled Russell's attempt to solve DirP of the MRTs.

But a worry remains. What is it for Iago to "take" panda fact (42) to be compared to the world in accordance with the rules (🐼)? If it is to have certain beliefs about fact (42) and rules (🐼), then circularity threatens. Now, in *NL*, 4–8, understanding is said to rest on "knowing" sense-stipulations, and both

[43] Which is why earlier I said that they are "elements" of belief.

the correlation of names with meanings and the "division" of worldly facts as like- or opposite-sense are characterized as "psychological." So it is tempting to think that some psychological states at least akin to belief is involved. Then the *NL* account is indeed circular.

It's not clear whether Wittgenstein had thought about this issue at the time of *NL*. But I want to offer a piece of speculation about how Wittgenstein might have sidestepped this circularity. Take "know" in *NL*, 4–8 to be Ryle's (1949) knowledge-how rather than propositional knowledge-that, and "psychological" in *NL*, 4–8 to point to whatever state of mind, distinct from propositional belief, that underlies this knowledge-how.[44]

4 A Wartime Notebook

Although the *NL* theory seems to resolve Russell's difficulty with the MRTs, Wittgenstein soon came to realize that it does *not* provide a satisfactory account of *falsity*. The realization came to him when he was in the Austrian army in World War I. During this time, he kept a number of notebooks, now published in *Notebooks, 1914–1916* (*NB*), in which he recorded his thoughts for a book on logic and philosophy – which of course was to be the *Tractatus*. In the first of these notebooks, in September 1914, Wittgenstein discusses something he calls the "truth-problem" (*Wahrheits-problem*). This problem has a number of aspects. I here focus on difficulties afflicting negative facts.

4.1 The Truth-Problem

On the *NL* theory, a conventional meaning for the English sentence

(48) Catiline denounced Cicero

may be captured by rules specifying a suitable fact about (48) as made false by a negative fact obtaining in the world.

But, intuitively, if (48) is false, it does *not* describe any aspect of the world. Doesn't this mean that there is *no fact* in the world to make (48) true? What, then, is this negative fact whose obtaining is of opposite sense to (48)?

Now, as mentioned at the end of Section 3.2, perhaps what persuaded Wittgenstein to accept negative facts is his conceiving them as absences of facts.

[44] Recently Stanley and Williamson (2001) argued against the existence of nonpropositional knowledge-how. But, in spite of criticizing Ryle, Stanley's (2011) account of "know-how" turns out to be in essentials Ryle's view.

In connection with my speculation, I note that some thirty years after *NL*, Wittgenstein would consider how our relations to rules might involve "customs (usages, institutions)," or "a practice" Wittgenstein (1997, Sections 199, 202).

So the obtaining of the negative fact making (48) false is the *absence of any fact* that corresponds to (48).

Now consider

(49) Iago loves Cassio

By parity of reasoning, what makes (49) false would then be the absence of any fact corresponding to it.

The absence of a fact, it would seem, is not a feature of the world; it is precisely not something obtaining in the world. So it's not clear how

1. the absence of any fact corresponding to (48)
2. the absence of any fact corresponding to (49)

are *different* absences.

But it seems these absences *have to be different*, because

- Absence 1 doesn't make (49) false.
- Absence 2 doesn't make (48) false.

Perhaps these absences are individuated by the constituents of the would-be facts involved. That is to say, it's because

- Absence 1 concerns Iago, *love*, and Cassio

while

- Absence 2 concerns Catilina, *denounce*, and Cicero,

that they are distinct absences. It's not clear, though, that this is satisfactory. After all, it is plausible that the sets

{Iago, *love*, Cassio}
{Catilina, *denounce*, Cicero}

exist. So, what is it for there to be *distinct nonexistences of facts* involving these entities?

These puzzles about the negative facts of *NL* is the source of Wittgenstein's agony:

> It is the *dualism*, positive and negative facts, that gives me no peace. For such a dualism can't exist. But how to get away from it? (*NB*, November 25, 1914)

This aspect of Wittgenstein's truth-problem is a version of an ancient problem of falsity: how is it possible to "say, speak, or think *that which is not* itself correctly by itself?" (Plato, *Sophista*, 238c).[45] Wittgenstein struggled with the problem of falsity throughout the first wartime notebooks. I turn to a brief look at some of these struggles.

4.2 The First Picture Theory and Common Form

One of the most famous aspects of the *Tractatus* is the *picture theory of propositions*. It is in trying to solve the problem of falsity that Wittgenstein hit upon the idea that propositions may be taken to be *pictures* or, more precisely, *models*:

> In the proposition a world is put together experimentally. (As when in the law-court in Paris an automobile accident is presented by means of dolls, etc.)
>
> This must yield the nature of truth straight away (if I were not blind). (*NB*, September 29, 1914)

Why does Wittgenstein think that reflecting on such models would yield him the nature of truth? To begin with, pictures "can be *right* and *wrong*" (*NB*, September 29, 1914). Moreover, it strikes Wittgenstein that when they're wrong, they represent something – say, the holding of a relation – that doesn't exist:

> A picture can represent relations that do not exist!!! (*NB*, September 30, 1914)

He immediately asks,

> How is that possible? (*NB*, September 30, 1914)

It is the evident fact that pictures represent the nonexistent that gives Wittgenstein the hope that, by reflecting on what makes this possible, he would reach a solution to the problem of falsity for propositions.

The first step Wittgenstein takes is to appeal to the idea of form from *NL*: a picture or proposition must have something in common with the world if it is to model the world, rightly or wrongly. Propositions presuppose that facts have a "logical structure":

> [I]n order for a proposition to be CAPABLE of having SENSE, the world must already have just the *logical structure* that it has. The logic of the world is prior to all truth and falsehood. (*NB*, October 18, 1914)

[45] See Narboux (2009) for an illuminating discussion of the relations between Plato's discussion of not-being and falsity in *Sophista* and Wittgenstein's concerns leading to the *Tractatus*.

> One must gather from the proposition the *logical structure of the facts* that makes it true or false. (*NB*, October 20, 1914; emphasis mine)

Pictures have "form," and this form is something in the picture that is "identical with reality":

> The *form of a picture* might be called that in which the picture MUST *agree with reality* (in order to be capable of portraying it at all). (*NB*, October 20, 1914; emphasis mine)
>
> The theory of logical picturing through language says – quite generally: In order for it to be possible that a proposition should be true or false – agree with reality or not – for this to be possible *something in the proposition* must be *identical with reality*. (*NB*, October 20, 1914; emphasis Wittgenstein's)

So the form of a proposition is identical with the logical structure of that fact that the proposition pictures.

It is far from clear what this common form is. But I want to set aside a discussion of difficulties with pinning down the idea of common form, and ask, how does common form help with the truth-problem?

On the *NL* view, a proposition *P* is made true by the obtaining of either a positive fact F^+ or by its negative opposite F^-. If *P* is made true by F^+, then it's made false by F^-; if made true by F^-, then made false by F^+. On each alternative, Wittgenstein thinks of the truth or falsity making fact as correlated with a proposition. The problem of falsity lies in there not being a coherent conception of negative fact to be correlated with a proposition as its truth or its falsity maker. According to the theory of propositions as pictures, what, in the first instance, corresponds in the world to a proposition is a structure or form, not a fact. No role is played by negative facts; F^- is replaced by absence of fact. No matter whether a picturing fact is made false or made true by the absence of a fact, these absences are individuated by would-be constituents. But, while sets of these constituents still exist, what does not exist is these constituents being combined in the particular way that constituents of the picturing fact are put together.

No sooner did Wittgenstein formulate this common-form solution to the truth-problem than he begins to doubt it:

> That in a certain sense the logical form of *p* must be present, even if *p* is not the case, shows itself symbolically through the fact that "*p*" occurs in "~ *p*."
>
> This is the difficulty: How can there be the form of *p*, if there is no fact of this form? And in that case, what does this form really consist in?! (*NB*, October 29, 1914)

The worry here has a Russellian source. For Russell, as we saw in Section 2.3.1, form or structure is some way in which entities are connected to constitute complexes. If a fact is absent, surely its would-be constituents are not connected. But then, a fortiori, there is no particular way in which they are connected, so there is no such thing as their form of connection to correspond to the false picture. (Form can't be like the grin of the Cheshire Cat, which remains after the Cat disappears.) The difficulty that this poses for the *NL* theory is immediate. According to the theory, the truth and falsity of a propositional fact *with a given form* is fixed by the existence of a positive or a *negative fact of the same form*. If the idea of negative facts having structure is incoherent, then the *NL* theory is incoherent.

4.3 The Emergence of Possibility

In early November 1914, Wittgenstein moves in another direction. He again brings up the now-familiar issue of falsity:

> How does the picture represent a fact?
>
> It is after all itself *not the fact*, which *need not be the case at all*. (*NB*, November 4, 1914; emphases mine)

Then he answers,

> One name represents one thing, another another thing, and they themselves are connected; thus the whole – like a *tableau vivant* – represents the fact. The logical connection must of course be *possible* for the represented things. (*NB*, November 4, 1914; first emphasis mine)

What in reality correspond to pictures are *not existing entities*:

- a fact, a connection of things, or a way in which things are connected

But rather *possibilities*:

- possible connections of things, or possible ways for the things to be connected.

This idea, and the truth-problem, are not taken up again in the notebooks. However, in the *Tractatus*, to which we now turn, *possibility* becomes central.

5 *Tractatus*

A number of doctrines from *NL* and *NB* reappear in the *Tractatus*:

- Propositions are facts.
- Propositions are pictures of facts.

- A proposition pictures a fact by having a form that the fact also has.
- The truth and falsity of propositions result from a comparison of propositional facts with the facts in the world that they picture.

However, the two main concepts appearing in these doctrines:

- Fact
- Form

are in the *Tractatus* based on a primitive notion of possibility. So in the *Tractatus* the meanings of these doctrines are transformed.

5.1 Fact, Object, State-of-Things

We start with the notion of fact from early in the *Tractatus*, in its famous opening:

1	The world is everything that is the case.
1.1	The world is the totality of facts [*Tatsachen*], not of things.
2	What is the case, the fact, is the existence of *states-of-things* [*Sachverhalte*].
2.01	A state-of-things is a *connection of objects (entities, things).*
2.03	In a state-of-things objects hang one in another, like the links of a chain.

2.03 is misleading. Intuitively, each link of an actual physical chain is an object in its own right, independent of being connected with the other links. But a Tractarian object isn't independent of states of things. Let's start with object:

2.011	It is essential to a thing that it can be a constituent part of a state-of-things.
2.0121(4)	Just as we cannot think of spatial objects at all apart from space, or temporal objects apart from time, so we *cannot think of any object apart from the possibility of its connection* with other things. If I can think of an object connected in a state-of-things, I cannot think of it apart from the *possibility* of this connection.

Objects are *not independent* of states-of-things, as the links of a chain are independent of that chain. Rather, essential to an object are its *possible occurrences* in state-of-things.

But states-of-things are equally not independent of objects:

2.0124 If all objects are given, then thereby are all *possible states-of-things* also given.

2.014 Objects contain the *possibility of all situations* [*Sachlage*].

As we will see in Section 6, what situations are possible is fixed by what states-of-things are possible. So both remarks claim that what states-of-things are possible depends on what objects there are – that is, on what *possibilities for combining into states-of-things* there are.

Wittgenstein tries to clarify this by an analogy with physical space:

2.013 Every thing is, as it were, in a space of possible states-of-things. I can think of this space as empty, but not of the thing without the space.

The idea that spatial objects are not thinkable apart from space alludes to a Kantian view that spatial objects are *essentially occupiers of spatial locations*, and so essentially spatially related to other spatial objects.

So a Tractarian object essentially "occupies" a possibility of connection with other objects. But in what sense does an object "occupy" a possibility of connection? Answer: the "occupation" of a possibility is the *realization* of that possibility.

A *possibility of connection* of objects is a *logical place*. Objects occupy that logical place just in case that possibility of connection is realized.

Here is a simplified model (following Sullivan 2001) of the space of possibilities as a two-dimensional plane with Cartesian coordinates. To begin with, think of each point on the x- and y-axes as an object. If the Cartesian coordinates of a point is (a, b), then this point models the *possibility* that the objects a and b are combined in a state-of-things.

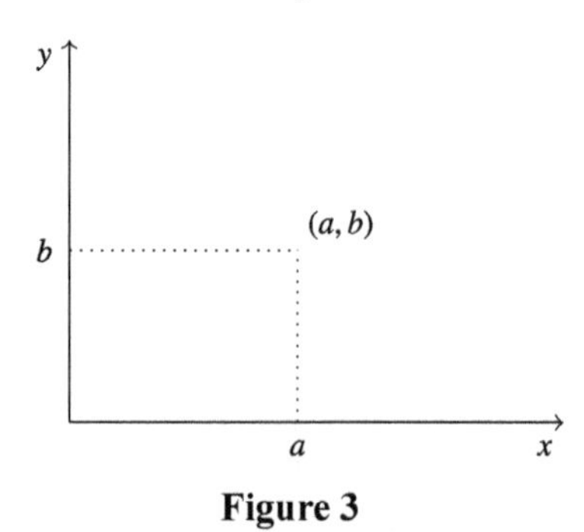

Figure 3

We can take point (a, b) in Figure 3 to be *blank* or not filled in, and to model a and b's *not* being combined in a state-of-things.

If, in contrast, (a, b) is *filled in*, as in Figure 4:

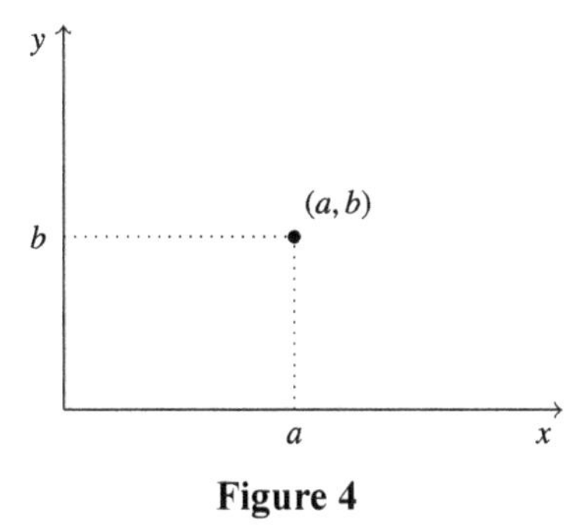

Figure 4

this models *a* and *b*'s being combined in a state-of-things.

In this model, an *object* can be thought of as a *vertical or horizontal line*:

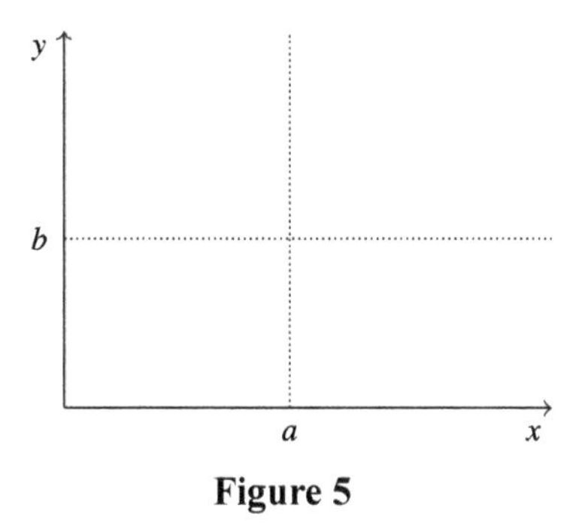

Figure 5

In Figure 5, the vertical line *is* the object *a*; the horizontal line *is* the object *b*.

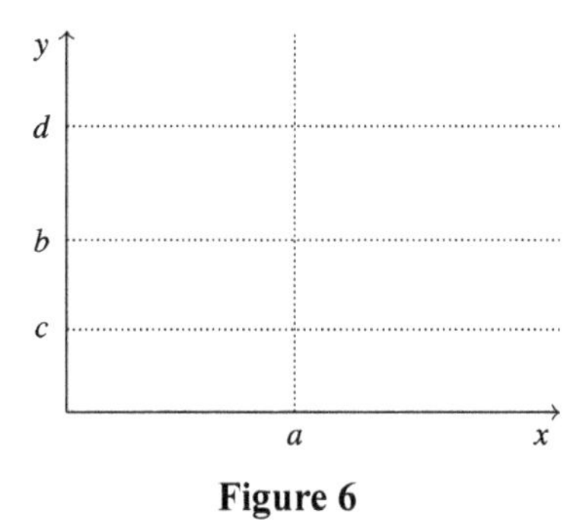

Figure 6

In Figure 6, along the vertical line that is *a* are all the points that can be occupied in virtue of *a*'s combining with some object on the *y*-axis – that is, all the possible combinations of *a* with objects. If the object *a* is the entire vertical line, then it *contains* all the points of its intersection with other objects.

Each point of intersection is a possible combination of *a* with that object. So, all possibilities of *a*'s occurring in a state-of-things are parts of *a*. It is *essential to a* that it intersects with all the horizontal lines. So it is essential to *a* that it contains all the possibilities of its combining with other objects. Possibilities of states-of-things involving finitely many objects are finite-dimensional spaces; objects are hyperplanes in such spaces.

In terms of this model, we may understand 2.03 as indicating that when objects are combined, there isn't, as it were, something, some relation, that holds them together, just as there isn't anything other than the links of a chain that hold the links together. In particular, Wittgenstein here rejects Russell's view that in complexes there is one relating relation that does the job of unifying the terms into that complex. Indeed, not only is there no further entity that unifies objects, but it is part of the very identity of each object that it can combine with objects.[46] Every object, one might say, is a kind of Russellian relating relation in that the unity of any state-of-things in which it occurs lies in its nature. We will soon see something of a rationale for this view.

The Russellian context for 2.03 allows us to clear up an exegetical debate over whether objects are all particular, or whether they include also properties and relations. We see now that there need be no real opposition between these views. What Wittgenstein is denying, in Russellian terminology, is that any object is a relating relation. We may take objects to include individuals and relations, so long as we abandon the Russellian view that some relation is *the* entity that unifies the remaining entities into a state-of-things.

5.2 Form Is the Possibility of Structure

Central to the *Tractatus* conception of proposition is a contrast between *structure* and *form* that's among the last ideas to make it into the book. The contrast appears in two remarks absent from the *Prototractatus* (Wittgenstein 1971), together with an addition to a remark in the *Prototractatus*. One *Prototractatus* absence is 2.033, setting out a general structure/form distinction:

2.031 In a state-of-things objects stand to one another in a determinate way.

2.032 The way in which objects hang together in a state-of-things is the *structure* of the state-of-things.

2.033 *Form* is the *possibility of structure*. (Emphasis mine)

[46] Wittgenstein's view that identity is not a dyadic relation perhaps precludes any object's possibly combining with itself into a state-of-things.

The remaining *Prototractatus* absences occur in the *Tractatus* discussion of picturing or modeling:

2.15 That the elements of the picture stand to one another in a determinate way represents [*stellt vor*] things as so standing to one another.

This connection of the elements of a picture is called its {structure, and the possibility of this structure its} form of depiction [*Form der Abbildung*].

2.151 The form of depiction is the possibility that the things stand to one another as do the elements of the picture.

2.151 isn't in *Prototractatus*. 2.15(1) is *Prototractatus* 2.151. The part of 2.15(2) in curly braces is an addition to *Prototractatus*:

2.15101 This connection of the elements of a picture is called its form of depiction.

Through adoption of the form/structure contrast, Wittgenstein takes the step merely contemplated in *NB* and incorporates possibility into picturing.

To understand this contrast, note two concepts of structure:

1. A structure may be the way in which a composite entity is put together, so that distinct composites can be put together in the same way. This is something like a pattern that can be instantiated by various entities.
2. One sometimes talks of a building as a structure. Here "structure" means a composite entity.

One *can* think of a picture or a state-of-things as a structure in the second sense – that is, as a composite entity. But in the *Tractatus*, the structure *of* a state-of-things is structure *in the first sense*. Two states-of-things, two connections of objects, can have the same structure if in each connection the objects stand to one another in the same determinate way.

A difficulty some may experience with the Tractarian structure/form distinction is that "form" is sometimes understood in Russellian fashion as the way in which some composite is put together. Conceived of in this way, it seems to make no sense to distinguish form from structure. Consider, however, Russell's own description of the problem posed by nonsymmetrical relations: "with a given relation and given terms, two complexes are 'logically possible'" (*TK*, 111). That is to say, he accepts that, prima facie, there are two *possible ways* in which a complex can be constituted from these ingredients. This means two possible complexes with *distinct possible structures*. For Russell, the notion of possibility has to be analyzed away; not so for Wittgenstein in the *Tractatus*. The notion of a *possible "way" of composing* is fundamental. Crucially, it is independent of whether there exists any composites realizing that "way." Wittgenstein's name for this notion is *form*.

The significance of the distinction is that a *state-of-things* is the *realization* or the *obtaining* of a *possibility*. A state-of-things obtains (*besteht*) just in case a possibility of things standing to one another in a determinate way is realized. It doesn't obtain if that possibility, that form of determinate connection, is not realized.

Let's consider how this conception resolves one crux of the truth-problem: "How can there be the form of *p*, if there is no fact of this form?" (*NB*, October 29, 1914). Russell doesn't distinguish form from structure, both are ways in which entities are connected to constitute complexes or facts. Hence a negative fact, an absence of connection of objects cannot coherently have a structure. The Tractarian insight is that in spite of this absence of structure, it is *possible* for objects to be connected in a specific way. There is form, even in the absence of structure.

Form brings Wittgenstein peace on positive and negative facts:

2.06(2) (The existence of states-of-things we also call a positive fact, their non-existence a negative fact.)

In the *Tractatus*, the notion of possibilities for things to connect into facts, and the notion of a possibility realized or not realized, obtaining or not obtaining, are primitive. Negative states-of-things are unrealized – that is, non-existent or non-obtaining – possibilities. This overcomes the puzzle of how they are individuated if understood as absences of facts. Negative facts are absences of facts, but absences whose specificities lie in different possibilities of structure.

Moreover, if objects are not in fact connected, then the ground, as it were, of the possibility of their combination into a state-of-things seems to have to lie in them. Each disconnected object must contain by itself the possibilities of connecting with others, as asserted in 2.014.

5.3 Picturing

We can now further unfold the view of propositions as pictures in the *Tractatus*. A "picture is a fact" (2.141). It is a fact "that [the] elements [of the picture] stand to one another in a determinate way" (2.14). These pictorial elements "correspond" to "objects" (2.13) and "are representatives, in the picture, of objects" (2.131). Statements 2.15–2.17 then provide an account of how these features of a picture underlie its being "a model of reality" (2.12).

Crucial to this account is the structure/form distinction for pictures set out in the 2.15s:

2.15(1) [T]he ... elements of a picture *stand to one another in a determinate way*

2.15(2) This *hanging together of the elements* of the picture is called its *structure* .... [T]he *possibility of this structure* is called its *form of depiction.*

Critical to what makes a picture represent is the structure it has:

2.15 The fact that the elements of a picture stand to one another in a determinate way represents that things stand so to one another.

This means that a pictorial fact "represents" that things stand to one another in the *same determinate way* as pictorial elements do. A picture represents this by means of the *form of depiction*:

2.151 The form of depiction is the *possibility* that things stand to one another as do the elements of the picture.

Picturing works like this. The pictorial elements stand to each other in a determinate way. This is what it is for a picture to be a fact; it is an *existing state* of pictorial elements. This state is a *realization of the possibility* that pictorial elements are connected in this determinate way, a realization of a form.

If this form is depicting form, then it is also *possible* for the things correlated with pictorial elements to be connected in a state-of-things with the *same* structure as the picturing fact. The possibility that pictured things stand to one another in this determinate way is the possibility that they are connected in a state-of-things. Thus, in virtue of this identity of possibility of connection, a picture *represents* a *possible* obtaining of a *state-of-things*:

2.201 The picture depicts reality by *representing* [*darstellt*] *a possibility* of the *obtaining ... of states-of-things*.

A propositional picture is true if the possibility it represents as obtaining is realized by the depicted objects, false otherwise. This conception makes room for false propositions as propositions representing possibilities not realized in the world. Thus, in the *Tractatus*, it is by the incorporation of modality into the nature of picturing that Wittgenstein finally solves the problem of falsity. So for Wittgenstein, the very notions of a proposition and of its truth or falsity are intrinsically modal.[47]

We have just seen that the truth and falsity of propositions are understood in terms of possibilities. So, in contrast to Frege and Russell, Wittgenstein does not conceive of possibility as a mode of truth. Possibility is primitive and prior to truth and falsity. It is only in virtue of representing possibilities that propositions are capable of being true or false. The primitiveness of possibility is of a piece with possibility's being "contrastive."[48] To conceive of possibility as a mode of truth is to think of possibility as what is the case elsewhere, say in some non-actual state of the world. Rejecting this conception of possibility means taking possibility to underlie both what is and what isn't the case. To borrow a turn of phrase from Frege, possibility is "poised between" (*schwanken*) what is and what isn't the case (Frege 1969, 8; Long and White translation Frege 1979, 7).

I pause briefly to take up a worry that NonP has crept back. A pictorial element is distinct from the object of which it's a representative. There's then no guarantee that they contain the same possibilities of combination with other objects. So that pictorial elements can combine into a state-of-things doesn't guarantee that the objects of which they are representatives can also combine into a state-of-things. A pictorial fact, then, may fail to represent any possibility, and so count as nonsense. This threat of nonsense presupposes that the correlation of object and representative can be set up independently of picturing, of the representative's combining into a fact whose possibility is of the obtaining of a fact involving the object. The *Tractatus* dodges the threat by insisting on a version of Frege's context principle:

3.33 only in the context of a proposition does a name have meaning.

To sum up, the conception of picturing in the *Tractatus* is not the same, but is a development of the common structure/form idea of *NB*. In the *NB* theory, the facts of *NL* are replaced with structure/form as the fundamental correlate in reality of propositions. In the *Tractatus*, Wittgenstein takes the final step merely considered in *NB*: form is an irreducible possibility. Thus, in the *Tractatus*, what

[47] I have set aside a complication stemming from Wittgenstein's realization that a picture – for example, of two people fencing (such as Wittgenstein's drawing in *NB*, September 29, 1914) – may be used to say that they are *not* fencing – that is, the possibility represented does *not* obtain (*NB*, October 30, 1914). In such a case of picturing falsity results from the people realizing the possibility of fencing.

[48] The term is Juliet Floyd's. For further discussion, see Floyd and Shieh (forthcoming).

is fundamental to pictures and propositions is *not* any part of reality, *but* rather the *possibilities for being part* of reality.

In this path to the *Tractatus*, both Russell's DirP and his notion of form play a role. The difficulties blocking a resolution of DirP lead to Wittgenstein's *NL* conception of (unified) propositional entities not vulnerable to the perplexities of false Moore–Russell propositions. The difficulties of deploying a Russellian conception of form in a coherent account of falsity for these *NL* propositions pointed Wittgenstein to the modal, non-Russellian forms of the *Tractatus*.[49] At the end of this road, modality has been brought back into the nature of propositions. Modality is now back in logic, to the extent that logic furnishes the standards of deductive validity, which is bounded by the truth or falsity of propositions. We will explore in more detail in the next section the connection between modality and logic in the *Tractatus*' rejection of Russell's conception of logic.

6 Logic and Logicism

A little more than half a year after Wittgenstein first went to Cambridge, he wrote to Russell that logic "must turn out to be a *totally* different kind than any other science" (*WC*, 30). The *Tractatus* is steadfast:

6.111 Theories which make a proposition of logic appear *substantial* are always false. [A proposition of logic] gets quite the character of a proposition of natural science and this is a certain symptom of its being falsely understood.

As late as 1918, Russell was unmoved by this strand of Wittgenstein's thinking, writing, "logic is concerned with the real world just as truly as zoology, though with its more abstract and general features" (1919a, 169). Reading the *Tractatus* in manuscript changed Russell's mind; on August 13, 1919, he wrote to Wittgenstein, "I am convinced you are right in your main contention, that logical prop[osition]s are tautologies, which are not true in the sense that substantial prop[osition]s are true" (*WC*, 96; see also *MPD*, 119).

In this section, we'll explore the differences between Russell's earlier conception of logic and that of the *Tractatus*.

[49] The Tractarian conceptions of proposition and form isn't free of difficulties and perplexities. Particularly puzzling is 2.172's seeming claim that a "picture cannot depict its form of depiction," but "shows it forth." In Shieh (forthcoming), I detailed an understanding of 2.172 based on the modal nature of Tractarian form, without, however, entirely dispelling the perplexity of 2.172.

6.1 The Propositions of Logic

We begin with an account of the propositions of logic in the *Tractatus*. The account rests on the picturing of language. This picturing has two components. At the bottom level of language are names, capable of combining into facts that are elementary propositions. These propositions depict states-of-things in the way elaborated in Section 5.3. Another level of language consists of nonelementary propositions. The world is the totality of states-of-things that obtain. Hence the world can be completely described by specifying which elementary propositions are true and which are false (4.26). A nonelementary proposition P represents in virtue of the representation of some set of elementary propositions into which P is uniquely "analyzed" (3.201, 3.25). Suppose P is analyzed into a finite set of n elementary propositions. There are 2^n possible ways for the n possible states-of-things represented by these elementary propositions to hold or fail to hold (4.27), and, correspondingly, 2^n ways for the n elementary propositions to be true or false (4.28). Each way for the n atomic propositions to be true or false is a "truth-possibility" of those elementary propositions (4.30). P then represents by expressing agreement and disagreement with each of these 2^n truth-possibilities. P is true if any of the truth-possibilities with which it agrees obtains, false otherwise. In this way each nonelementary proposition is a truth-function of elementary propositions, and each elementary proposition is truth-function of itself. But one and the same truth function of elementary propositions can be expressed in different ways, in which different "logical constant" signs occur. So the logical signs are not representatives of any objects; no objects play any role in how propositions represent (4.0312).

A proposition corresponds to a class of truth-possibilities, those with which it agrees; these are the proposition's truth-conditions (4.41). Since there are 2^n truth-possibilities for n elementary propositions, and each proposition analyzable into these elementary ones either agrees or disagrees with each of these truth-possibilities (4.42), there are 2^{2^n} possible distinct propositions analyzable into these elementary ones. Most important, one of these agrees with every truth-possibility and another disagrees with every truth-possibility. These propositions are, respectively, a tautology and a contradiction (4.46). It seems clear, then, that a tautology is true no matter which elementary propositions are true, no matter what states-of-things obtain, no matter how the world is. It is not made true by correctly picturing the world. The same holds for contradictions with falsity in place of truth. For such reasons tautologies and contradictions "say nothing," have "no truth-conditions," are "senseless" (4.461), "not pictures of reality," and "do not represent any possible situation" (4.462). The last characterization shows that situations are what nonelementary propositions depict.

Now, if tautologies have no truth-conditions, it's not altogether clear that they *are* true; nor is it clear that contradictions *are* false.[50] Still, it might seem that we can now specify why, according to Wittgenstein, the propositions of logic are fundamentally different from those of natural science. The propositions of natural science are made true or false by correctly or incorrectly picturing the world. In contrast, if tautologies are true at all, it is not in virtue of being correct representations, but in virtue of *how* they represent the world; similarly, the falsity of contradictions arises from the mechanism of propositional representation rather than the world.

This characterization may well make it seem that Wittgenstein's disagreement with Russell is over the ground of the truth of propositions of logic, not over whether logic consists of a special set of propositions, true in some distinct way.

One reason to be suspicious of such a view of Wittgenstein's Tractarian conception of logic is

6.122 [W]e can even [*auch*] do without logical propositions.

I now show that in the *Tractatus*, logic is not fundamentally a special set of truths at all.

6.2 The Picturing of Truth-Functions

We begin with a puzzle about the distinction between elementary and nonelementary propositions in the *Tractatus*.

The modal conception of picturing as presenting a possibility for things applies to elementary propositions. Nonelementary propositions are characterized in two ways: as we just saw, they are truth-functions of elementary propositions, but, in addition, since they are propositions, they are "pictures of reality" (4.01). It is clear that for a proposition p to be a truth-function of other propositions is for the truth or falsity of p to be determined entirely by the truth or falsity of the other propositions. However, this characterization does not answer the question of how p counts as a picture.

Consider, for example, the following explanation of why a car wouldn't start:

> Either the battery is dead or there's no gas in the tank.

The truth-value of this disjunctive proposition is obviously fixed by the truth-values of its two disjuncts. But what situation does this proposition depict? It doesn't describe

[50] See Dreben and Floyd (1991) for further discussion of this issue.

- The situation in which the battery is dead.
- Nor the situation in which the tank is empty.
- Nor the situation in which both the battery is dead and the tank is empty.

Any one of these three situations could fail to obtain without rendering the proposition false.

The answer to this puzzle lies in the modal nature of picturing. A truth-function counts as a picture because, like an elementary proposition, it also presents a possibility, but not the same kind of possibility as presented by an elementary proposition. Consider, for example, a disjunction "*p* or *q*" of elementary propositions, which is a truth-function of *p* and of *q*. Each disjunct presents a possibility for things, and each of these possibilities can be either realized or not realized. What the disjunction affirms is that one of three *possibilities* is the case:

- that in which *p* is true but *q* is false,
- that in which *q* is true but *p* is false,
- or that in which both are true.

That is to say, a disjunction pictures these three possibilities as open to being realized. In addition, the disjunction also

- denies that the remaining possibility, in which neither *p* nor *q* is true, is open to being realized,

that is, it pictures this possibility as closed to realization.

Now, *p* is true if the possibility it presents is realized and false if that possibility is not realized, similarly for *q*. Thus, a disjunction presents a possibility for how the possibilities presented by its disjuncts are realized or not realized in the world. Since picturing is the presentation of possibilities, it follows that disjunctions, and truth-functional propositions in general, are pictures.

On this view of the picturing of disjunctions, a disjunction isn't a fully distinct picture from its disjuncts. A disjunction pictures possibilities for the possibilities presented by its disjuncts, so the disjunction pictures by means of the possibilities pictured by its conjuncts. Wittgenstein takes the disjunctive proposition to be *internally related* to the disjuncts (for discussion of "internal relation" see Proops 2002).

From the internal relation between a disjunction and its disjuncts flows certain necessities and impossibilities. For example, if *R* is the disjunction "*P* or *Q*":

- It's impossible for *P* to picture correctly at the same time that *R* pictures incorrectly.
- If *R* pictures correctly, then it is necessary that at least one of *P* and *Q* pictures correctly as well.

These necessities and impossibilities govern the picturing of a disjunction and the picturing of its disjuncts. They constitute necessary patterns of norms governing the picturing of these propositions. These impossibilities and necessities are also impossibilities and necessities of the truth and falsity of the propositions *P*, *Q*, and *R*. So we can think of them equally as certain impossible and necessary patterns of truth-values of *P*, *Q*, and *R*. Their necessity lies in:

- It is not intelligible for *P* to picture correctly at the same time that "*P* or *Q*" pictures incorrectly.
- It makes no sense to take "*P* or *Q*" to picture correctly and yet at the same time to take neither *P* nor *Q* to picture correctly.

Such patterns are, I claim, what Wittgenstein means by "rules of logical syntax" (3.334).[51]

Tractarian patterns of picturing underlie logical "relations" of *implication* and *incompatibility*. For example, the patterns just displayed underlie:

- *P* implies *R*
- The falsity of both *P* and *Q* is incompatible with the truth of *R*

These implications and incompatibilities are founded on the nature of nonelementary picturing, and the picturing of these propositions *P*, *Q*, and *R*. They would not be the propositions they are without these implications and incompatibilities among them.

6.3 The Nature of Logic

The foregoing account of nonelementary picturing points to the nature of logic.

Logical syntax underlies implications and incompatibilities among propositions. These implications and incompatibilities are the standards of correctness

[51] For extended discussion and defense of this reading of "logical syntax" in the *Tractatus*, see Shieh (2014, 2015, forthcoming). I hold, in particular, that, in addition to rules of logical syntax such as that governing disjunction, there are rules governing the symbolizing of single propositional-signs such as: it is not intelligible for one part of such a sign to function logically as a proper name and the remaining part to function as a second-level concept-expression. The attribution of the latter sort of rules to the *Tractatus* is tacitly acknowledged by resolute interpretation such as Conant (2002) and Diamond (1991).

governing all deductive reasoning that logic comprises. Thus, logic lies in the natures of the propositions involved in reasoning. They do not rest in *other*, logical, propositions.

Logic is not concentrated in a special set of propositions, but pervades all propositions. This is not to deny that there are such things as propositions of logic or that these propositions are tautologies. It is, rather, to claim that the propositions of logic are of peripheral interest in logic, which explain 6.122's insistence that we have no need of propositions of logic.

What is essential to logic, then, is what pictures of reality there can be, which in turn determines how propositions can be internally related to one another. That is to say, the essence of logic consists of "the most general form of propositions," which are all possibilities of elementary and nonelementary picturing, "every possible sense" (4.5).

If logic lies in all nonelementary picturing, then we are, as it were, always already in the grip of logic, if we think at all, if we picture the world at all. Whence

5.4731 [L]anguage itself prevents every logical mistake. – That logic is a priori consists in this: that nothing illogical *can* be thought.

6.4 A Sign-Language with Operator N

In 4.5, Wittgenstein alludes to a "specification" of the most general propositional form by "a description of the propositions of some one sign-language" suitable for expressing every possible sense. One reason for seeking such a specification is that logical features and "relations" are *not obvious* from expressions of propositions in "everyday language"; it is "humanly impossible to gather immediately from it the logic of language" 4.002(3). Rewriting "everyday" propositions in a "sign-language" in which all possible senses are expressible allows ordinary, "material," inferential "relations" to be seen to rest on their (inexplicit) logical forms.

In the 5s, Wittgenstein sketches such a sign-language. It has only one logical operator, written "N," which expresses joint negation of all propositions to which it is applied. " N" generalizes a Sheffer stroke (see Sheffer 1913). In this language,

6.001 every proposition is a result of the successive application of the operation $N'(\bar{\xi})$ to elementary propositions.

This "is the general form of propositions" (6).

N is attached to an "indication" of the multiplicity or class of propositions to which it is applied, and in 5.501, Wittgenstein gives three ways of describing such a multiplicity of propositions:

1. Direct enumeration. ...
2. Specification of a function fx, whose values for all values of x are the described propositions.
3. Specification of a *formal law*, according to which ... propositions are constructed. In this case the [described propositions] are all the terms of a *form-series*.

A language with such specifications of operands matches some of the expressive capabilities of the language(s) of *PM*.

6.5 First-Order Logic

The propositional functions of description 2 result from "*turning*" a "constituent part" of a proposition into a *variable* (3.315). This means *substituting* a variable for that constituent. (See, e.g., Goldfarb 2018, 176.)

A *value* of a propositional function ϕ is a proposition resulting from substituting, for the variables in ϕ, legitimate or suitable constituents of propositions. The values constitute the range of the propositional function. Legitimate propositional constituents are arguments to or of the propositional function ϕ.

Let call the results of substituting variables for *names* occurring in propositions "(first-order propositional) functions of name variables." Applications of the *N* operator to description 2 functions of names express first-order quantifications. Wittgenstein gives an example:

5.52 If the values of ξ are the totality of values of a function fx for all values of x, then $\mathrm{N}(\overline{\xi}) = \sim (\exists x).fx$.

"N" negates all values of fx – that is, it says that no value of fx holds. Negating this says some value of fx holds. Once description 2 is augmented with notational devices for indicating scopes of quantifications, the language suffices for expressing first-order logic (see, inter alia, Floyd 2001).

Wittgenstein holds that identity is not a relation among objects, so that identity and difference should not be expressed by a dyadic predicate but by identity and difference of signs. Hintikka (1956) first formulated methods of implementing this together with first-order generalization. See Rogers and Wehmeier (2012) for a definitive account of Wittgensteinian first-order logic with identity.

6.6 Higher-Order Logic

The logical system of *PM*, in service of Russell's logicism, involves higher-order generalizations over predicates and relations. This brings up the question for Wittgenstein's description 2: What "constituents" of a proposition *other than names* can be converted into variables?

One proposal is that (atomic) predicates, those that occur in elementary propositions, can be converted into variables, to obtain second-order propositional functions of predicate variables with predicates as arguments. (Compare, e.g., Potter 2009, 270; Goldfarb 2018, 177.)

A critical question is then:

> What are the values of such propositional functions, and what are their arguments, what can be legitimately substituted for a predicate variable?

We could take first-order propositional functions to be logically complex predicates. Then the question is:

> Which first-order propositional functions, if any, are legitimately substitutable for the predicate variables of a second-order propositional function?

A constraint on answering these questions is the Tractarian view that a proposition constructed by logical operations on others is a truth-function of those other propositions. Thus a second-order quantification such as

$$\forall X(Xa \supset Xb) \qquad (※)$$

is a truth-function of the values of the propositional function

$$\lambda X(Xa \supset Xb) \qquad (✠)$$

and these values are the instances of the quantification (※). Hence the truth-values of all the propositions in the range of propositional function (✠) have to be determined independently of and prior to the truth-value of quantification (※). This is a predicativity constraint on legitimate arguments to propositional functions of predicate variables.

In order to satisfy this constraint, no argument to a second-order propositional function

$$\lambda X(\Phi(X))$$

of the predicate variable $X\xi$ can be a first-order function that contains quantifications formed from this very second-order function $\lambda X(\Phi(X))$. Tractarian quantification with predicate variables has to conform to this version of Russell's vicious-circle principle.

This Tractarian vicious-circle constraint is clearly not enough for a complete account of legitimate first-order arguments to second-order propositional functions. In particular, it leaves open the question whether something like Russell's ramified hierarchy of propositional functions of individuals is consistent with the constraint. It's not clear whether Wittgenstein's claim that Russell's axiom of reducibility is redundant helps with this question. Instead of pursuing this matter, I turn to description 3, which embodies a still relatively little-studied idea in logic.[52]

6.7 (Anti-)Logicism

Description 3 provides an expression of the ancestral of a relation. In 4.1252 and 4.1273, Wittgenstein presents a form-series of propositions

$$aRb,\ (\exists x)(aRx\ \&\ xRb),\ (\exists x)(\exists y)(aRx\ \&\ xRy\ \&\ yRb),\ \ldots \qquad (\dagger)$$

Taking R to be the relation of being a parent of, these propositions say, in order, a is a parent of b, a is a grandparent of b, a is a great-grandparent of b, and so forth. Their joint negation then says that a is not an ancestor of b, so its negation says a is an ancestor of b. Negating the application of N to † expresses what is known as the ancestral of the relation R.

The significance of this is that Russell's (and Frege's) arithmetical logicism may be factored into two parts. One is the definition of numbers as classes of equinumerous classes. The other is the definition of the number zero and the ancestral of a relation of immediate successor in order to pick out the natural numbers and prove the Peano axioms. The definition of the ancestral requires either impredicative second-order quantification or the axiom of reducibility.

Expression of the ancestral of R using (†) apparently dispenses with higher-order or class-theoretic quantification, and so perhaps affords some initial support for

6.031 The theory of classes is completely superfluous in mathematics.

Merely initial support for a couple of reasons. First, Wittgenstein never tells us what is the "formal law" for "constructing" this form-series (†). In fact, he never explains what a "formal law" is. Second, we have no idea how this expression of the ancestral leads to arithmetical truths.

[52] Ricketts (2013, 134) argues that not even first-order functions in which only first-order quantifiers occur are legitimate arguments to predicate variables. In Shieh (forthcoming), I raise doubts about this claim on the basis of a(n infinitary) procedure for constructing Tractarian propositions conforming to description 2.

Let's start with the second issue. Note that Wittgenstein in fact rejects logicism in Russellian or Fregean form; he does not hold that arithmetical truths are provable from basic logical truths and definitions. Indeed, Wittgenstein holds that Frege's and Russell's way of expressing the ancestral is "false; it contains a *circulus vitiosus*" (4.1273). Unfortunately he provides no explanation of this vicious circle.

Wittgenstein would certainly also regard Frege's conception of numbers as logical objects with suspicion, given his insistence that logical constants are not representatives. It's controversial whether Russell took numbers to be logical entities in *PM*. Some interpreters, such as Church (1976), Goldfarb (1989), Hylton (1990), and Quine (1961), take Russell's characterization of it as a "no-classes" theory to mean that classes are eliminated in favor of propositional functions. If propositional functions are entities, then on such interpretations, numbers are logical entities. Other interpreters, such as Klement (2010) and Landini (1998), hold that in *PM*, quantification over propositional functions is substitutional, and there is no commitment to these entities. The latter seems to go with Russell's later claim that "classes are merely a convenience in discourse," and the propositional functions in favor of which classes are eliminated are "only ... expression[s]" (*PMD*, 62). It's unclear that Wittgenstein would object to such a substitutional version of Russellian logicism.

Wittgenstein doesn't reject all connection of mathematics and logic; rather,

6.2 Mathematics is a logical method.
The propositions of mathematics are equations – that is, apparent propositions.

Numbers are not (logical) objects; rather

6.021 A number is the exponent of an operation.

This brings us back to the first issue. There's a range of interpretations of Wittgenstein's "formal law": inter alia Geach (1981, at 170), Potter (2009, at 272), Ricketts (2013), Fisher and McCarty (2016), and Weiss (2017).

At one end of the range is Weiss's very minimalist conception: a formal law specifies the *substitution* of propositions into a schema that results from a given by replacing one or more occurrences of its component propositions with a placeholder. At the other end is Fisher and McCarty's maximalist conception: a formal law is a primitive recursive function from propositions to propositions in an arithmetical coding.

Even with Weiss's minimalist form-series there is a reconstruction of the ancestral of 4.1252 and 4.1273. More interesting, Weiss in effect reconstructs

6.021's "exponent of an operation" in terms of an idea I call "exponents of a relation." If *a* is a parent of *b*, then think of them as a related by one application of the *parent of* relation; if a grandparent, then two applications of *parent of*, and so on. Numbers lie in iterations of applications of a relation *R*. Using this idea, Weiss formulates a logical system incorporating minimalist form-series that affords the means of mimicking first-order arithmetic.

Is this a form of logicism?

This is a large and complex question on which I can here offer only the briefest comment.

Substitutional Russellian logic is not committed to logical objects such as truth-functions, quantifiers, or propositional functions. It may seem little different in conception from Tractarian logic. Perhaps that explains Russell's coming to accept, in 1919, that logical truths are "not true in the sense that substantial propositions are true." Substitutional Russellian logicism would then hold that, in being derived from logic, arithmetic also comprises a set of such special truths.

However, it's not clear whether the mimicking of arithmetic in the *Tractatus* consists in the derivation of arithmetical truths from tautologies. Thus, in contrast to substitutional Russellian logicism, the *Tractatus* takes no stand on whether arithmetic and logic are non-substantial in the same way.

For much more extensive discussion of logic and anti-logicism in the *Tractatus* than I can attempt here, see Floyd (2001, 2021), Frascolla (1994, 1997, 2005), Goldfarb (2018), Kremer (2002), Landini (2007, 2021b), and Marion (1998) as well as Juliet Floyd's Element in this series.

7 After Multiple-Relation Theories

In Wittgenstein's manuscripts after the *Tractatus*, there are over 150 mentions of Russell. In this section, I discuss briefly Russell's theory of belief after he finally rejected MRTs and Wittgenstein's critique of it at the beginning of the "middle" period of his philosophy, since the critique connects with the Tractarian conception of proposition.[53] I end by mentioning two well-known issues on which Wittgenstein's later thinking takes off from Russell.

7.1 Russell on Belief after Multiple-Relation Theories

As we saw, from 1913 to 1918, Russell didn't accept the *NL* (re)conception of propositions as facts in a medium of representation. He was left, in *PLA*, with

[53] My discussion here is deeply indebted to Engelmann (2018, manuscript). For illuminating discussions of Wittgenstein's "middle" philosophy, see Stern (1995, 2007).

the "two-verb" MRT according to which the object-relation has the puzzling feature of both relating and not relating. While in prison in 1918 for anti-war activities shortly after delivering the *PLA* lectures, Russell read Watson's behaviorist psychology and began formulating a new theory of belief. The theory appears in "On Propositions" (*OP*) and *The Analysis of Mind* (Russell 1921, hereafter *AM*). In these works, Russell finally rejects MRTs and brings in propositional facts, constituted from mental images or words, with "image-propositions" "more primitive" (*OP*, 29; *AM*, 241–2).

Russellians are fond of pointing out that Russell's stated reason for abandoning MRTs is they are "rendered impossible by the rejection of the subject" (*OP*, 27), and inferring therefrom that Russell's rejection of MRTs couldn't have anything to do with Crit_1 or Crit_2. This reading is plausible, but not mandatory. To begin with, on Russell's account, the "most important thing about" the new conception is that a proposition "is, whenever it occurs, an actual fact, having a certain analogy of structure … with the fact which makes it true or false" (*OP*, 30). Russell's new theory is a less general version of *NL*: propositions are facts whose constituents are images or words, and are made true or false by presence or absence of worldly facts with "analogous" structures. Moreover, Russell holds that an advantage of the new theory is "making it possible to admit propositions as actual complex occurrences, and doing away with the difficulty of answering the question: what do we believe we believe falsely?" (*OP*, 27). What makes the two-verb MRT puzzling is that a "subordinate verb" such as *love* somehow has to occur as relating Desdemona and Cassio without implying that "non-existent love between Desdemona and Cassio" (*PLA*, 59). By taking a proposition to be an actual fact involving images, Russell no longer has to countenance the peculiarities of either second verbs lurking in judgment complexes or nonexistent facts. Moreover, in *OP*, the reason Russell gives for rejecting the subject is:

> The … subject[is] not empirically discoverable. It seems to serve the same sort of purpose as is served … by numbers and particles and the rest of the apparatus of mathematics. All these things have to be constructed, not postulated …. The same seems to be true of the subject. (*OP*, 25)

Indeed, in 1918, Russell had already sketched a logical construction of a person as "a certain series of experiences" (*PLA*, 374). It's not clear why a variant MRT could not be based on such a construction: a belief state is a complex constituted by a multiple relation relating the items from which a subject is logically constructed and the objects of the belief. Thus, the available evidence doesn't preclude Wittgenstein's having influenced Russell's new fact-based theory of belief.

However,

5.631(2) in an important sense, there is no subject
5.632 The subject does not belong to the world but is a limit of the world.

most likely did *not* influence Russell's rejection of the subject.

Image-propositions are only one component of Russell's theory of beliefs in *AM*. An image-proposition is the common "content" of a variety of kinds of beliefs, kinds of attitudes to the same propositional content: expectation, memory, hope, fear, and so forth. The attitudes are distinguished by feelings or "emotions" (*AM*, 208) associated with a propositional content (*AM*, 205).

One kind of belief, expectation-belief, is central to Russell's account of "all verification" of beliefs, including verification of "scientific hypothes[e]s (*AM*, 270). Verification "consists in the happening of something expected" and "the experience of verification" occurs when "accustomed activities have results that are not surprising" (*AM*, 269; see also Russell 1927, 189) Russell advances an "external and causal view of the relation of expectation to expected occurrence":

> We have first an expectation, then a sensation with the feeling of expectedness related to memory of the expectation. This whole experience, when it occurs, may be defined as verification, and as constituting the truth of the expectation. (*AM*, 270)

An expectation is "falsified" "when a situation arises which gives a feeling of … surprise [in connection with it]" (*AM*, 269)

One focus of discussion of Wittgenstein's *Philosophical Remarks* and lectures in the 1930s is this Russellian external causal theory of expectation. Wittgenstein notes that Russell's theory is avowedly based on behaviorist accounts of mental phenomena such as wishing on a simple model of hunger: "Russell treats wish (expectation) and hunger as if they are on the same level" (1980, 9). Russell takes hunger to be a "behavioral-cycle" (*AM*, 65). At the beginning of the cycle is an "animal which is hungry," a "restless" or uncomfortable state (*AM*, 62). This state causes the behaviors of searching for food. If the search succeeds, the animal eats and, if it is enough food, the animal experiences satisfaction that halts its searching behavior (*AM*, 62–3). "The 'purpose' of a behavioral-cycle is the result that brings it to an end" (*AM*, 65), so food is the purpose correlated to hunger; it is the object of the desire or wishing that is hunger.

Against the general applicability of this behavioral cycle model to desire, Wittgenstein tells a well-known joke:

> If I wanted to eat an apple and someone punched me in the stomach, taking away my appetite, then it was this punch [according to Russell's theory] that I originally wanted. (1975, Section 22)

A similar objection faces a conception of commands on this behavioral model:

> if I command something to someone, and if what he does makes me happy (it gives me a feeling of expectedness), then according to Russell I should conclude that he performed the command (1975, Section 22)

The moral that these difficulties point to is that while "several things will satisfy my hunger, my wish (expectation) can only be fulfilled by something definite" (1980, 9).

It should be clear that parallel issues arise for falsification of expectations. Suppose you expect that Dora will arrive at 5. She does, but wearing clothes strikingly different from what she usually wears. Suppose you then experience a feeling of surprise, not expectedness. Does that falsify your expectation? So it's not merely the verification of wish or expectation that requires something definite; falsification does as well. This is connected to the problem of determinate falsity of judgment in *NB*. Not just any absence from the world falsifies a given proposition, and Russell's feeling of surprise, by itself at any rate, is no adequate alternative to the *Tractatus*' modal determinations of propositional falsity.

7.2 Two More Themes

The so-called private language argument is a much-discussed part of *Philosophical Investigations*. Some salient features of Wittgenstein's exchanges with his interlocutors on privacy have Russellian echoes. In "Knowledge by Acquaintance and Knowledge by Description," Russell asserts that, if "there is such a thing as direct acquaintance with oneself," only Bismark is directly acquainted with himself, and can use "Bismark" as a name, as opposed to a definite description, of himself (*KAKD*, 114). So, for Bismark, "Bismark" in "some statement" (*KAKD*, 114) "refer[s] to what can only be known to" Bismark (Wittgenstein 1997, Section 243).[54]

After 1919, Russell explicitly connects these ideas with privacy. He claims that "data obtained by introspection are private and only verifiable by one observer"(*OP*, 12); "images remain private in a sense in which sensations are not" (*AM*, 110). He also argues for the existence of images:

[54] Diamond (2000) argues that an argument against private language is tacit in the *Tractatus*.

> If you try to persuade an ordinary uneducated person that she cannot call up a visual picture of a friend sitting in a chair, but can only use words describing what such an occurrence would be like, she will conclude that you are mad. (*OP*, 11)

Wittgenstein asks, "In what sense are my sensations private?" (Wittgenstein 1997, Section 246). He wonders, "What does it mean when we say: … 'What would it be like, if it were otherwise?' … , when someone has said that my images are private" (Wittgenstein 1997, Section 251). And he suggests, "If you say he sees a private picture before him, which he is describing, you have still made an assumption about what he has before him. And that means that you can describe it" (Wittgenstein 1997, Section 294).

One of the many intriguing ideas in *Remarks on the Foundations of Mathematics* is the "surveyability" of mathematical proof discussed in Part III. A substantial part of the discussion is directed at the "Russellian calculus" of *PM*. The concern is, yet again, Russellian (and Fregean) logicism. Wittgenstein sketches something like a thought-experiment: two patterns of proof in *PM* have such a large number of quantifiers and bound variables that we cannot tell whether they are the same pattern without counting the quantifiers and variables using our everyday Arabic numeral notation. Is a *PM* derivation then really the logicist foundation of our everyday practices with numbers? Wittgenstein isn't rejecting logicism but rather reconceiving what logicism comes to. He thereby explores the possibility of avoiding psychologism's loss of objectivity with an "anthropological" and yet not subjective perspective on mathematics and logic. (For extensive discussion of *RFM*, III, see Floyd 2022; Mühlhölzer 2005, 2009.)

It would be a fine irony if, from such a perspective, mathematics appears to be "non-extensional" (Floyd 2021, 5, passim), since it is a significant aspect of Russell's reception of the *Tractatus* that the second edition of *PM* embraces extensionality. (For further discussion, see Linsky 2011).[55]

[55] It should be noted that the extensionality that Russell takes from the *Tractatus* is the thesis that every proposition is a truth-function of elementary propositions, which is compatible with the Tractarian conception of possibility as what underlies what is and isn't the case, rather than a mode of truth.

A Closing Word

The friendship between Wittgenstein and Russell waned after the *Tractatus*. Russell had harsh words for Wittgenstein's later philosophy:

> I have not found in Wittgenstein's *Philosophical Investigations* anything that seemed to me interesting .... [T]he later Wittgenstein ... seems to have grown tired of serious thinking and to have invented a doctrine which would make such an activity unnecessary. (*MPD*, 161)

Wittgenstein is no kinder in return, for he wrote to Maurice Drury:

> Russell's books should be bound in two colours ... those dealing with mathematical logic in red – and all students of philosophy should read them; those dealing with ethics and politics in blue – and no one should be allowed to read them. (Rhees 1984, 112)

And, in a letter to Moore dated December 3, 1946, Wittgenstein described Russell's participation at the Cambridge Moral Science Club as "[g]lib and superficial" (*WC*, 405).

It should be noted, however, that Wittgenstein has no binding recommendations for Russell's writings in philosophy of mind and language. Moreover, in the same 1946 letter to Moore, Wittgenstein wrote that Russell was "as always, *astonishingly* quick" (*WC*, 405). Indeed, Norman Malcolm noted that at the Moral Science Club meetings, "Wittgenstein was deferential to Russell in the discussion as I never knew him to be with anyone else" (Malcolm 1958, 57).

If philosophy as Wittgenstein conceives of it is "responsive" (Diamond 2018, 281), then, as we have seen, responses to Russell's "line[s] of thinking" (Wittgenstein 1998, 16) are central in Wittgenstein's philosophy, starting from his path to the *Tractatus* and continuing throughout the twists and turns of his later reflections.

Abbreviations

Wittgenstein

NB *Notebooks, 1914–1916*
NL "Notes on Logic"
RFM *Remarks on the Foundations of Mathematics*
WC *Wittgenstein in Cambridge. Letters and Documents 1911–1951*

Russell

AM *The Analysis of Mind*
KAKD "Knowledge by Acquaintance and Knowledge by Description"
MPD *My Philosophical Development*
NTF "On the Nature of Truth and Falsehood"
OP "On Propositions"
PLA "The Philosophy of Logical Atomism"
PM *Principia Mathematica*
PoM *The Principles of Mathematics*
PoP *The Problems of Philosophy*
TK *Theory of Knowledge*

Moore

NJ "The Nature of Judgment"

Other

Crit_1 Wittgenstein's criticism of Russell on May 20, 1913
Crit_2 Wittgenstein's criticism of Russell on May 26, 1913
DirP The direction problem
DRT Dual-relation theory
MRT Multiple-relation theory
NonP The nonsense problem

PB	Permutative belief
Prob_{Bel}	The problem of belief individuation
$\text{Prob}_{\text{Coord}}$	The problem of coordination between beliefs and truth-conditions
Prob_{TC}	The problem of truth-conditions individuation
TPB	The theory of permutative beliefs

Bibliography

Aristotle (1964). *Analytica Priora.* Ed. W. D. Ross. Clarendon.

Bostock, D. (2012). *Russell's Logical Atomism.* Oxford University Press.

Brentano, F. C. (1874). *Psychologie vom empirischen Standpunkt*, vol. 1. Duncker & Humblot.

Carey, R. (2007). *Russell and Wittgenstein on the Nature of Judgement.* Continuum.

Church, A. (1976). "Comparison of Russell's Resolution of the Semantical Antinomies with That of Tarski". *Journal of Symbolic Logic* 41(4): 747–60.

Conant, J. (2002). "The Method of the *Tractatus.*" In Reck 2002, 227–51.

Connelly, J. (2011). "On Props, Wittgenstein's June 1913 Letter, and Russell's Paralysis." *Russell: The Journal of Bertrand Russell Studies* 31(2): 141–66.

(2014). "Russell and Wittgenstein on Logical Form and Judgement." *Theoria* 80(3): 232–54.

(2021). *Wittgenstein's Critique of Russell's Multiple Relation Theory of Judgement.* Anthem.

Davidson, D. (1969). "On Saying That", In *Inquiries into Truth and Interpretation.* Clarendon, 93–108.

Diamond, C. (1991). "What Nonsense Might Be," In *The Realistic Spirit.* MIT Press, 95–114.

(2000). "Does Bismarck Have a Beetle in His Box?" In A. Crary and R. Read, editors, *The New Wittgenstein.* Routledge, 262–92.

(2018). "Commentary on José Zalabardo's 'the *Tractatus* on Unity.'" *Australasian Philosophical Review* 2(3): 272–84.

Dreben, B., and J. Floyd (1991). "Tautology: How Not to Use a Word." *Synthese* 87: 23–49.

Dummett, M. (1973). *Frege: Philosophy of Language.* Duckworth.

Elkind, L. D. C., and G. Landini, eds. (2018). *The Philosophy of Logical Atomism.* Springer.

Engelmann, M. L. (2018). "Phenomenology in Grammar." In O. Kuusela, M. Ometităa, and T. Uçan, eds. *Wittgenstein and Phenomenology.* Routledge, 22–46.

(manuscript). *Wittgenstein's* Philosophical Remarks.

Ferber, M. (2019). *Poetry and Language.* Cambridge University Press.

Fine, K. (2000). "Neutral Relations." *Philosophical Review* 109(1): 1–33.

Fisher, D. and C. McCarty (2016). "Reconstructing a Logic from *Tractatus*." In S. Costreie, ed. *Early Analytic Philosophy*. Springer, 301–24.

Floyd, J. (2001). "Number and Ascriptions of Number in Wittgenstein's *Tractatus*." In J. Floyd and S. Shieh, eds. *Future Pasts*. Oxford University Press, 145–91.

(2021). *Wittgenstein's Philosophy of Mathematics*. Cambridge University Press.

(2022). " 'Surveyability' in Hilbert, Wittgenstein and Turing." *Philosophies* 7.

Floyd, J. and A. Kanamori (2016). "Gödel vis-á-vis Russell." In G. Crocco and E. M. Engelen, eds. *Kurt Gödel*. Presses Universitaires de Provence, 243–376.

Floyd, J. and S. Shieh (forthcoming). "Modality in Wittgenstein's *Tractatus*." In J. Zalabardo, ed. *The* Tractatus: *A Critical Guide*. Cambridge University Press.

Frascolla, P. (1994). *Wittgenstein's Philosophy of Mathematics*. Routledge.

(1997). "The *Tractatus* System of Arithmetic." *Synthese* 112(3): 353–78.

(2005). *Understanding Wittgenstein's* Tractatus. Routledge.

Frege, G. (1879). *Begriffsschrift*. L. Nebert.

(1969). *Nachgelassene Schriften*. Felix Meiner.

(1979). *Posthumous Writings*. Ed. H. Hermes, F. Kambartel, and F. Kaulbach. Trans. P. Long and R. White. Blackwell.

Geach, P. T. (1981). "Wittgenstein's Operator N." *Analysis* 41(4): 168–71.

Giaretta, P. (1997). "Analysis and Logical Form in Russell: The 1913 Paradigm." *Dialectica* 51(4): 273–93.

Goldfarb, W. D. (1989). "Russell's Reasons for Ramification." In Savage and Anderson 1989, 24–40.

(2018). "Wittgenstein against Logicism." In E. H. Reck, ed. Logic, *Philosophy of Mathematics, and Their History*. College Publications, 171–83.

Griffin, N. (1985). "Russell's Multiple Relation Theory of Judgment." *Philosophical Studies* 47(2): 213–48.

(1991). "Was Russell Shot or Did He Fall?" *Dialogue* 30(4): 549–54.

Halimi, B. (2013). "Structured Variables." *Philosophia Mathematica* 21(2): 220–46.

Hanks, P. W. (2007). "How Wittgenstein Defeated Russell's Multiple Relation Theory of Judgment." *Synthese* 154(1): 121–46.

(2015). *Propositional Content*. Oxford University Press.

Hintikka, J. (1956). "Identity, Variables, and Impredicative Definitions." *Journal of Symbolic Logic* 21: 225–45.

Hochberg, H. (1976). "Russell's Attack on Frege's Theory of Meaning." *Philosophia* 18: 9–34.

Hylton, P. (1990). *Russell, Idealism, and the Emergence of Analytic Philosophy*. Oxford University Press.

Johnston, C. (2012). "Russell, Wittgenstein, and Synthesis in Thought." In J. Zalabardo, ed. *Wittgenstein's Early Philosophy*. Oxford University Press, 15–36.

Jubien, M. (2001). "Propositions and the Objects of Thought." *Philosophical Studies* 104(1): 47–62.

Klement, K. C. (2004). "Putting Form before Function: Logical Grammar in Frege, Russell, and Wittgenstein." *Philosopher's Imprint* 4(4): 1–47.

(2010). "The Functions of Russell's No-Class Theory." *Review of Symbolic Logic* 3(4): 633–64.

Kremer, M. (2002). "Mathematics and Meaning in *Tractatus*." *Philosophical Investigations* 25(3): 272–303.

Landini, G. (1991). "A New Interpretation of Russell's Multiple-Relation Theory of Judgment." *History and Philosophy of Logic*, 37–69.

(1998). *Russell's Hidden Substitutional Theory*. Oxford University Press.

(2007). *Wittgenstein's Apprenticeship with Russell*. Cambridge University Press.

(2014). "Types* and Russellian Facts." In B. Linsky and D. Wishon, eds. *Essays on* The Problems of Philosophy. Center for the Study of Language and Information, 233–75.

(2021a). *Repairing Bertrand Russell's 1913 Theory of Knowledge*. Macmillan.

(2021b). "Tractarian Logicism." *Review of Symbolic Logic* 14(1): 973–1010.

Lebens, S. (2017). *Bertrand Russell and the Nature of Propositions*. Routledge.

Levine, J. (2004). "On the 'Gray's Elegy' Argument and Its Bearing on Frege's Theory of Sense." *Philosophy and Phenomenological Research* 69: 251–95.

(2013). "*Principia Mathematica*, the Multiple-Relation Theory of Judgment and Molecular Facts." In N. Griffin and B. Linsky, eds. *The Palgrave Centenary Companion to Principia Mathematica*. Macmillan, 247–304.

Linsky, B. (1999). *Russell's Metaphysical Logic*. Center for the Study of Language and Information.

(2011). *The Evolution of* Principia Mathematica. Cambridge University Press.

(2018). "The Near Riot over Negative Facts." In Elkind and Landini 2018, 181–97.

MacBride, F. (2007). "Neutral Relations Revisited." *Dialectica* 61(1): 25–56.
(2013). "The Russell–Wittgenstein Dispute: A New Perspective." In M. Textor, ed. *Judgement and Truth in Early Analytic Philosophy and Phenomenology*. Macmillan, 206–41.
Malcolm, N. (1958). *Ludwig Wittgenstein*. Oxford University Press.
Marion, M. (1998). *Wittgenstein, Finitism, and the Foundations of Mathematics*. Oxford University Press.
McGuinness, B. (1988). *Wittgenstein, a Life*. University of California Press.
McTaggart, J. M. E. (1908). "The Unreality of Time." *Mind* 17(68): 457–74.
Misak, C. (2020). *Frank Ramsey*. Oxford University Press.
Moltmann, F. (2003). "Propositional Attitudes without Propositions." *Synthese* 135: 70–118.
Monk, R. (1990). *Ludwig Wittgenstein*. Free Press.
Moore, G. E. (1899). "The Nature of Judgment." *Mind* 8(30): 176–93. (*NJ*)
(1901). "Truth and Falsity." In J. M. Baldwin, ed. *Dictionary of Philosophy and Psychology*. Vol. 1. Macmillan, 716–18.
Mühlhölzer, F. (2005). " 'A Mathematical Proof Must Be Surveyable.' " *Grazer Philosophische Studien* 71: 57–86.
(2009). *Braucht die Mathematik eine Grundlegung*? Klostermann.
Narboux, J. P. (2009). "Négation et totalité dans le *Tractatus* de Wittgenstein." In C. Chauviré, ed. *Lire le* Tractatus *de Wittgenstein*. Vrin, 127–76.
Noonan, H. W. (2001). *Frege*. Polity.
Pears, D. F. (1977). "The Relation between Wittgenstein's Picture Theory of Propositionsand Russell's Theories of Judgment." *Philosophical Review* 177–96.
(1989). "Russell's 1913 *Theory of Knowledge* Manuscript." In Savage and Anderson 1989, 169–82.
Perovic, K. (2018). "Can We Be Positive about Russell's Negative Facts?" In Elkind and Landini 2018, 199–218.
Pincock, C. (2008). "Russell's Last (and Best) Multiple-Relation Theory of Judgement." *Mind* 117(465): 107–39.
Plato (1985). *Sophista*. Vol. 1: *Platonis Opera*. Ed. J. Burnet. Clarendon.
Potter, M. D. (2009). "The Logic of the *Tractatus*." In D. M. Gabbay and J. Woods, eds. *Handbook of the History of Logic*. Vol. 5. Elsevier, 255–304.
Proops, I. (2002). "The *Tractatus* on Inference and Entailment." In Reck 2002, 283–307.
(2011). "Russell on Substitutivity and the Abandonment of Propositions." *Philosophical Review* 120(2): 151–205.
Quine, W. V. (1961)."Logic and the Reificationof Universals." In *From a Logical Point of View*. 2nd ed. Harvard University Press, 102–29.

Reck, E. H., ed. (2002). *From Frege to Wittgenstein: Perspectives on Early Analytic Philosophy*. Oxford University Press.

Rhees, R., ed. (1984). *Recollections of Wittgenstein: Hermine Wittgenstein, Fania Pascal, F. R. Leavis, John King, M. O'C. Drury*. Oxford University Press.

Ricketts, T. (1996). "Pictures, Logic, and the Limits of Sense in Wittgenstein's *Tractatus*." In H. Sluga and D. G. Stern, eds. *The Cambridge Companion to Wittgenstein*. Cambridge University Press, 59–99.

(2013). "Logical Segmentation and Generality in Wittgenstein's *Tractatus*." In M. D. Potter and P. M. Sullivan, eds. *Wittgenstein's* Tractatus. Oxford University Press.

Rogers, B. and K. F. Wehmeier (2012). "Tractarian First-Order Logic: Identity and the N-operator." *Review of Symbolic Logic* 5(4): 538–73.

Rumfitt, I. (1994). "Frege's Theory of Predication: An Elaboration and Defense, with Some New Applications." *Philosophical Review*, 599–637.

Russell, B. (1903). *The Principles of Mathematics*. Cambridge University Press. (*PoM*).

(1904). "Meinong's Theory of Complexes and Assumptions (III)." *Mind* 13(52): 509–24.

(1905a). "Necessity and Possibility." In *The Collected Papers of Bertrand Russell*. Vol. 4. Ed. A. Urquhart and A. C. Lewis. Routledge, 507–20.

(1905b). "On Denoting." *Mind* 14(56): 479–93.

(1906a). "On Substitution." In *The Collected Papers of Bertrand Russell*. Vol. 5. Ed. G. H. Moore. Routledge, 129–232.

(1906b). "On the Nature of Truth." *Proceedings of the Aristotelian Society* 7: 28–49.

(1910–11). "Knowledge by Acquaintance and Knowledge by Description." *Proceedings of the Aristotelian Society* 11: 108–28. (*KAKD*).

(1910). "On the Nature of Truth and Falsehood." In *Philosophical Essays*. Longmans, Green, 147–59. (*NTF*).

(1912a). *The Problems of Philosophy*. Williams and Norgate. (*PoP*).

(1912b). "What Is Logic?" In *The Collected Papers of Bertrand Russell*. Vol. 6. Ed. J. G. Slater and B. Frohmann. Routledge, 55–6.

(1913a). "Props." In Russell 1984, 195–9.

(1913b). *Theory of Knowledge*. In Russell 1984, 1–178. (*TK*).

(1918–19). "The Philosophy of Logical Atomism." *The Monist*. (*PLA*).

(1919a). *Introduction to Mathematical Philosophy*. George Allen and Unwin.

(1919b). "On Propositions." *Proceedings of the Aristotelian Society Supplementary Volume*, (2), 1–43. (*OP*).

(1921). *The Analysis of Mind.* Allen and Unwin. (*AM*).

(1927). *An Outline of Philosophy*. Allen and Unwin.

(1944). "Replies to Criticisms." In P. A. Schilpp, ed. *The Philosophy of Bertrand Russell*. Open Court, 680–741.

(1959). *My Philosophical Development*. Allen and Unwin. (*MPD*).

(1984). *The Collected Papers of Bertrand Russell.* Vol. 7: *Theory of Knowledge: The 1913 Manuscript*. Ed. E. R. Eames and K. Blackwell. George Allen & Unwin.

Ryle, G. (1949). *The Concept of Mind.* University of Chicago Press.

Savage, C. W. and C. A. Anderson, eds. (1989). *Rereading Russell.* University of Minnesota Press.

Van der Schaar, M. (2013). *G. F. Stout and the Psychological Origins of Analytic Philosophy*. Palgrave Macmillan.

Sheffer, H. M. (1913)."A Set of Five Independent Postulates for Boolean Algebras, with Application to Logical Constants." *Transactions of the American Mathematical Society* 14(4): 481–8.

Shieh, S. (2012). "Logic, Modality, and Metaphysics in Early Analytic Philosophy." In L. Haaparanta and H. Koskinen, eds. *Categories of Being: Essays on Metaphysics and Logic*. Oxford University Press, 293–318.

(2014). "In What Way Does Logic Involve Necessity?" *Philosophical Topics* 42(2): 289–337.

(2015). "How Rare Is Chairman Mao?" In B. Weiss, ed. *Dummett on Analytical Philosophy*. Macmillan, 84–121.

(2017). "Pragmatism, Apriority, and Modality: C. I. Lewis against Russell's Material Implication." In C. Sachs and P. Olen, eds. *Pragmatism in Transition*. Palgrave Macmillan.

(2019). *Modality and Logic in Early Analytic Philosophy*. Vol.1: *Necessity Lost.* Oxford University Press.

(2021). "Strict Implication and the Pragmatic A Priori." In Q. Kammer, J. P. Narboux, and H. Wagner, eds. *C. I. Lewis*. Routledge, 104–31.

(2022). "On Attempting to Solve the Direction Problem." *Russell: The Journal of Bertrand Russell Studies* 42: 132–68.

(forthcoming). *Modality and Logic in Early Analytic Philosophy*. Vol. 2: *Necessity Regained.* Oxford University Press.

Soames, S. (2014). *The Analytic Tradition in Philosophy*. Vol. 1: *The Founding Giants*. Princeton University Press.

(2017). *The Analytic Tradition in Philosophy*. Vol. 2: *A New Vision.* Princeton University Press.

Sommerville, S. (1981). "Wittgenstein to Russell (July 1913): 'I Am Very Sorry to Hear ... My Objection Paralyses You.' " In R. Haller and

W. Grassl, eds. *Proceedings of the 4th International Wittgenstein Symposium*. Holder-Pichler-Tempsky, 182–8.

Stanley, J. (2011). *Know How*. Oxford University Press.

Stanley, J. and T. Williamson (2001). “Knowing How.” *Journal of Philosophy* 98(8): 411–44.

Stern, D. G. (1995). *Wittgenstein on Mind and Language*. Oxford University Press.

(2007).“Wittgenstein,the Vienna Circle, and Physicalism, a Reassessment.” In A. Richardson and T. Uebel, eds. *The Cambridge Companion to Logical Empiricism*. Cambridge University Press, 305–31.

Stevens, G. (2003). “Re-examining Russell’s Paralysis: Ramified Type-Theory and Wittgenstein’s Objection to Russell’s Theory of Judgment.” *Russell: The Journal of Bertrand Russell Studies* 23(1): 5–26.

(2005). *The Russellian Origins of Analytical Philosophy*. Routledge.

Stout, G. F. (1910). “The Object of Thought and Real Being.” *Proceedings of the Aristotelian Society* 11: 187–205.

Sullivan, P. M. (2001). “A Version of the Picture Theory.” In W. Vossenkuhl, ed. *Ludwig Wittgenstein: Tractatus Logico-Philosophicus*. Akademie, 89–110.

Turnau, P. (1991). “Russell’s Argument against Frege’s Sense-Reference Distinction.” *Russell: The Journal of Bertrand Russell Studies* 11: 52–66.

Weiss, M. (2017). “Logic in the *Tractatus*.” *Review of Symbolic Logic* 10(1): 1–50.

Whitehead, A. N. and B. Russell (1910). *Principia Mathematica*. 1st ed. Vol. 1. Cambridge University Press. (*PM*).

Wittgenstein, L. (1913). “Notes on Logic.” In *Notebooks, 1914–1916*. Trans. G. E. M. Anscombe. 2nd ed. Blackwell, 93–107. (*NL*).

(1922). *Tractatus Logico-Philosophicus*. Trans. F. P. Ramsey. Kegan Paul, Trench, Trubner & Company.

(1971). *Prototractatus*. Cornell University Press.

(1975). *Philosophical Remarks*. Trans. R. White and R. Hargreaves. University of Chicago Press.

(1978). *Remarks on the Foundations of Mathematics*. Trans. G. E. M. Anscombe. revised. MIT Press. (*RFM*).

(1979). *Notebooks, 1914–1916*. Ed. G. H. von Wright and G. E. M. Anscombe. Trans. G. E. M. Anscombe. 2nd ed. Blackwell. (*NB*).

(1980). *Lectures, Cambridge 1930–1932*. Ed. D. Lee. Basil Blackwell.

(1997). *Philosophical Investigations*. Trans. G. E. M. Anscombe.Blackwell.

(1998). *Culture and Value*. 2nd ed. Blackwell.

(2008). *Wittgenstein in Cambridge*. Ed. B. McGuinness. Blackwell. (*WC*).

Acknowledgments

Thanks to David Stern for inviting me to contribute to this Elements series. I've been greatly helped by discussions with Bernie Linsky, Rosalind Carey, Mauro Engelmann, and Jim Levine. I'm grateful to Kevin Klement for extensive helpful comments. Thanks to two referees for incisive comments and helpful suggestions; apologies for being prevented by word limits from identifying specific improvements they occasioned. Thanks to my students in a seminar on the metaphysics of thought for reading and responding to an earlier version. I owe, as always, a great philosophical debt to Juliet Floyd for many illuminating conversations over the years on Russell's MRTs and the *Tractatus*.

The Philosophy of Ludwig Wittgenstein

David G. Stern
University of Iowa

David G. Stern is a Professor of Philosophy and a Collegiate Fellow in the College of Liberal Arts and Sciences at the University of Iowa. His research interests include history of analytic philosophy, philosophy of language, philosophy of mind, and philosophy of science. He is the author of *Wittgenstein's Philosophical Investigations: An Introduction* (Cambridge University Press, 2004) and *Wittgenstein on Mind and Language* (Oxford University Press, 1995), as well as more than 50 journal articles and book chapters. He is the editor of *Wittgenstein in the 1930s: Between the 'Tractatus' and the 'Investigations'* (Cambridge University Press, 2018) and is also a co-editor of the *Cambridge Companion to Wittgenstein* (Cambridge University Press, 2nd edition, 2018), *Wittgenstein: Lectures, Cambridge 1930–1933, from the Notes of G. E. Moore* (Cambridge University Press, 2016) and *Wittgenstein Reads Weininger* (Cambridge University Press, 2004).

About the Series

This series provides concise and structured introductions to all the central topics in the philosophy of Ludwig Wittgenstein. The Elements are written by distinguished senior scholars and bright junior scholars with relevant expertise, producing balanced and comprehensive coverage of the full range of Wittgenstein's thought.

Cambridge Elements

The Philosophy of Ludwig Wittgenstein

Elements in the Series

Wittgenstein on Aspect Perception
Avner Baz

Reading Wittgenstein's Tractatus
Mauro Luiz Engelmann

Wittgenstein on Logic and Philosophical Method
Oskari Kuusela

Wittgenstein on Sense and Grammar
Silver Bronzo

Wittgenstein on Forms of Life
Anna Boncompagni

Wittgenstein on Criteria and Practices
Lars Hertzberg

Wittgenstein on Religious Belief
Genia Schönbaumsfeld

Wittgenstein and Aesthetics
Hanne Appelqvist

Style, Method and Philosophy in Wittgenstein
Alois Pichler

Wittgenstein on Realism and Idealism
David R. Cerbone

Wittgenstein and Ethics
Anne-Marie Søndergaard Christensen

Wittgenstein and Russell
Sanford Shieh

A full series listing is available at: www.cambridge.org/EPLW

For EU product safety concerns, contact us at Calle de José Abascal, 56–1°,
28003 Madrid, Spain or eugpsr@cambridge.org.

www.ingramcontent.com/pod-product-compliance
Ingram Content Group UK Ltd.
Pitfield, Milton Keynes, MK11 3LW, UK
UKHW020306140625
459647UK00006B/60